ROCK MECHANICS

Other books published within the Series on Rock and Soil Mechanics

Reisner, W. & v. Eisenhart Rothe, M.:
Bins and Bunkers
for Handling Bulk Materials
1971

Dreyer, W.:
The Science of Rock Mechanics
Part I: Strength Properties of Rocks
1972

Hanna, T. H.:
Foundation Instrumentation
1973

Gregory, C. E.:
Explosives
for North American Engineers
First Edition
1973

Reimbert, M. & A.:
Retaining Walls Volume I
Anchorages and Sheet Piling
1974

Vutukuri, V. S. & Lama, R. D.:
Handbook on Mechanical Properties
of Rocks
Volume I
1974

Reimbert, M. & A.:
Retaining Walls Volume II
Study of Passive Resistance
in Foundation Structures
1976

Hardy, H. R., Jr. & Leighton, F. W.:
First Conference on Acoustic
Emission / Microseismic Activity
in Geologic Structures and Materials
1977

Karafiath, L. L. & Nowatzki, E. A.:
Soil Mechanics
for Off-Road Vehicle Engineering
1977

Baguelin, F., Jézéquel, J. F.
& Shields, D. H.:
The Pressuremeter
and Foundation Engineering
1978

Lama, R. D. & Vutukuri, V. S.:
Handbook on Mechanical Properties
of Rocks
Volumes II, III & IV
1978

Gregory, C. E.:
Explosives
for North American Engineers
Second Edition
1979

Assonyi, Cs. & Richter, R.:
The Continuum Theory
of Rock Mechanics
1979

Hardy, H. R., Jr. & Leighton, F. W.:
Second Conference on Acoustic
Emission / Microseismic Activity
in Geologic Structures and Materials
1979

Editor-in-Chief
Professor Dr.-Ing. H. Wöhlbier

Series on Rock and Soil Mechanics
Vol. 3 (1978/79) No. 5

ROCK MECHANICS

by

Alfreds R. Jumikis
Professor of Civil Engineering
Rutgers University
The State University of New Jersey
USA

First Edition
1979

TRANS TECH PUBLICATIONS

Distributed in North America by
TRANS TECH PUBLICATIONS
16 Bearskin Neck, Rockport, MA 01966, USA

and world-wide by
TRANS TECH S. A.
CH-4711 Aedermannsdorf, Switzerland

ISBN 0-87849-026-4

Printed in the United States of America

PRINTED AND BOUND IN THE UNITED STATES OF AMERICA

FOREWORD

In this volume, the author presents an introductory segment to the relatively new civil engineering discipline known as engineering rock mechanics. This subject is presented here from the viewpoint of a civil engineer to civil engineers.

The content of this book deals with rock as an engineering construction material by means of which, upon which, and within which civil engineers build structures in rock. The discipline thus pertains to hydraulic structures engineering; to highway, railway, canal, foundation, and tunnel engineering; as well as to earthworks of, and substructures in, rock of all kinds in any way associated with engineering.

The main purpose of this book is to assist interested readers in understanding some of the basic rock mechanics principles as they apply to rock engineering. Hence, the book is developed basically as a guide in engineering rock mechanics.

Nowadays, rock mechanics develops very rapidly. The recent output of books, conference proceedings, research reports, and various articles on rock mechanics may be regarded as evidence of realization of the importance of the rock mechanics discipline in geotechnical engineering.

It should be said at this point that presently there are many published references and comprehensive discourses about various aspects of rock mechanics already available. As a matter of fact, rock mechanics literature indicates that such recently published material is literally soaked with new approaches in rock mechanics. However, in many instances some of the publications are not easy for a reader with limited time and/or knowledge to extract the basic information he needs; nor is it easy for him to evaluate and compare research papers and analyze the encountered engineering problems in rock. Hence, there is a need for abridging this condition. In other words, there is a need for a basic reading material such as the one presented in this book from which to learn about the nature of rock as an engineering construction material. These circumstances prompted the author to prepare this modest volume.

This book is written in simple language. It is easy to read. The basic concepts presented are easy to grasp and understand. Nevertheless, the text of the book is explicit. It concentrates on essentials, thus enabling the reader to obtain a certain amount of adequate knowledge for understanding and comprehending some basic principles, methods, and their use in rock mechanics, as well as the performance of rock under the influences of load, water, and temperature.

In essence, this unique volume emphasizes understanding. It gives a practical orientation to basic rock mechanics; provides a background as well as an outlook that motivates to further study; and will allow the reader to profit from his later studies of more comprehensive and complex publications on engineering rock mechanics than what is presented in this text.

May this book prove enjoyable and profitable to the reader.

ALFREDS R. JUMIKIS

College of Engineering
Piscataway, New Jersey

Christmas 1978

TABLE OF CONTENTS

Part 2
ROCK EXPLORATION

Chapter 4
METHODS OF ROCK EXPLORATION

TECHNOLOGICAL PROPERTIES OF ROCKS

IN-SITU STRENGTH OF ROCKS

Chapter 5
ROCK MASS PROPERTIES

Part 3
STRESSES IN ROCK ABOUT CIRCULAR UNDERGROUND OPENINGS

Chapter 6
STRESS FIELDS

Chapter 7
ELASTIC STRESS ANALYSIS IN ROCK ABOUT UNDERGROUND OPENINGS

Chapter 8
PLASTIC ZONES IN ROCK AROUND UNDERGROUND OPENINGS

Chapter 9
STRESSES IN ELASTIC ROCK AROUND VERTICAL SHAFTS

Chapter 10
SOME ENGINEERING PROBLEMS ASSOCIATED WITH WORK IN ROCK

Part 4
STABILIZATION OF ROCK

Chapter 11
ROCK REINFORCEMENT

Chapter 12
ROCK SLOPES

APPENDICES

PREFACE

The various aspects of engineering rock mechanics are concerned with engineering design and construction in rock.

In the study of engineering rock mechanics, rock is considered to be a construction material by means of which, upon which, and in which engineers build structures.

Generally, the geotechnical engineer must know the physical, mechanical, and strength properties of the materials with which he works.

From the civil engineering point of view, the increasing magnitude in size and scope of engineering projects designed and built in rock, and the subsequent resulting increase in responsibility to produce such projects, demand from the engineer qualitative as well as quantitative knowledge about rock as a construction material.

Rocks loaded by a structure may undergo displacements and deformations or, if overloaded, they may become damaged by cracking or crushing. Hence, the effects of loads on rock strata depend on the physical properties of such materials. In any subsurface excavation and construction, the strength properties of rocks must be known.

Although a commonly used construction material, rock is a very complex one. It is very difficult to describe, and most difficult to define. Hence, the study of rocks in engineering rock mechanics is an important branch of civil engineering in general, and in foundation engineering, hydraulic structures engineering, and subsurface construction engineering in particular.

In designing structural foundations in rock, engineers encounter problems which are frequently of a geological nature. It is so because all structures are laid on the ground.

Because geological and environmental factors expose difficulties when working in rock and, thus, influence the cost of construction, one's familiarization with the knowledge of physical geology as well as with the principles of engineering geology is helpful as well as profitable in engineering rock mechanics and rock engineering practice. However, where geology serves the purpose of construction, geology must be subordinated to the laws of engineering mechanics.

The basic idea in design of an engineering structure on or in rock is that the structure and the structure-supporting rock *en masse* as an integral unit should be considered as an inseparable, integral, functional and statical system, and that in this sense, every statical analysis of such rock structure systems should conform to and bear with this situation, and should be so dealt with.

This is so because the engineer has to accept the rock in which he builds as it occurs in its natural state with all of its own weaknesses, characterized by stratification, jointing, fisssures, faults and folds, variable rock properties, weathering, and the like. Of course, the strength properties of the rock may be better with increase in depth than at/or near the ground surface. Or else the strength of the rock may be increased by chemical injections and grouting, thus increasing the stability of the structure. These latter measures, however, are usually limited by some economical and/or technical restrictions.

If proper consideration is to be given to the most economical design and safe performance of a structure on or in rock, adequate knowledge about the engineering properties of the subsurface rocks must be available. Thus, because of todays large construction projects in rock, the engineer resorts to the theory of rock mechanics.

Although there is no theory of strength of rocks available as yet, in the context of rock mechanics, the latter may be in a way considered as the youngest branch of the subject of strength of materials dealing with rock as a construction material.

To conclude, it is apropos to say that the period of empiricism in handling rocks is now over. In its place, there enter scientific and new engineering concepts. The basic prerequisite for designing and constructing in rock, besides geological structures, stratigraphy and other pertinent information, is the knowledge of rock as an engineering material.

ACKNOWLEDGMENTS

The author expresses his cordial thanks to all those who contributed intellectually and materially, in words and in deeds, to the preparation of this manuscript for publication.

For tangible furtherance of this work, the author expresses his appreciation to Dr. E. H. Dill, Dean, College of Engineering, Rutgers, The State University of New Jersey, New Brunswick, New Jersey.

For providing secretarial services for typing this manuscript, the author expresses his sincere thanks to Dr. R. C. Ahlert, Executive Director of the Bureau of Engineering Research, and to Dr. J. Wiesenfeld, Chairman, Department of Civil and Environmental Engineering, both at Rutgers University.

The author is especially grateful to Dr. Evert Hoek for his furnishing to the author his great works on Rock Slope Engineering, and Underground Excavation Engineering, and for his constructive suggestions.

The author also acknowledges Dr. Edward J. Cording's helpful written discussion about rock wedges formed in a circular sidewall of an underground opening.

The author is also most appreciative of the excellent library service of the Rutgers Library of Science and Medicine. Its efforts in procuring the many reference sources and books permitted personal examination and verification of facts in original, bona fide references.

To the authors credited in the documentation of this book, the writer owes a great deal, because some of their ideas and work helped in building up this work. Credit to these and other authors is given through references in the text as extensively as possible.

In addition to sources cited in the body of the text, the author acknowledges the many publications made available to him by various persons and organizations. Permissions to reproduce their published material were granted by the following individuals:

Dr. ARTHUR CASAGRANDE, Harvard University
Dr E. CORDING, Professor, University of Illinois
Dr. D. U. DEERE, formerly Professor in the University of Illinois
Dr. E. HOEK, formerly Professor in the Imperial College of Science and Technology, London
Dr. E. R. LEEMAN, Assistant Director, National Mechanical Engineering Research Institute, South African Council for Scientific and Industrial Research, Pretoria, South Africa
Dr. M. ROCHA, President, Council of Civil Engineering Research, Lisbon, Portugal.

The author is also glad to acknowledge the following publishers for the permissions granted for him to use their material:
Plenum Publishing Corporation, New York
McGraw-Hill Book Company, New York
E. and F. N. Spon Ltd., London
Springer Verlag, Berlin
John Wiley and Sons, Inc., New York.

This book has been prepared within the course of the author's academic, research, and practice activities relating to rocks and engineering rock mechanics and, in a way, is the outgrowth of a set of lecture notes prepared for a course entitled "Engineering Rock Mechanics," taught by the author in the College of Engineering at Rutgers, The State University of New Jersey.

A separate word of thanks is also expressed to the author's family. The author's son, ANDRIS ALFREDS JUMIKIS, P. E. in Civil Engineering, participated in research about the engineering properties of Triassic shale; helped in laboratory testing of granites, gneisses, and marbles; prepared a number of pencil sketches and all line drawings for this book; and assisted the author in his field work. The author's wife, ZELMA ALBERTINE JUMIKIS, deserves special credit for her patience in organizing and maintaining the voluminous files of notes pertaining to this work, and for her never-tiring moral support for this project.

Special thanks are conveyed herewith to Mrs. ERMA M. SUTTON for editing the author's handwritten draft of the manuscript for this book and for carefully typing and processing the manuscript. Her close attention to the work greatly expedited its preparation.

Likewise, appreciation is expressed to Mrs. Doris Clark for her patient typing of parts of the manuscript.

The publisher has lived up to its best tradition, and the author's cordial thanks go to all concerned in the production of this book.

A. R. J.

PART 1

THE SUBJECT

CHAPTER 1

INTRODUCTION

1-1. Some Historical Notes

For the purpose of one's orientation, let us avail ourselves of a brief historical review about the development of rock mechanics, but more specifically of rock mechanics technology.

Since prehistoric times, the various strength properties of rocks have been recognized by primitive man. The caveman was aware of the stability of the roof of his cave when choosing his shelter. The Stone-Age man tried to choose optimum places to strike a piece of rock when carving tools and making flint modules for weapons. Also, by necessity, it was of interest throughout the history of human civilization to understand the performance characteristics of rock when subject to loads.

Likewise, the art of subsurface excavation has been known to mankind for many centuries. Subsurface mining, first in pits and later through shafts, began during the Stone Age some 15,000 years ago.

It is reported (12) that tunneling started about 3,500 B.C. during the Bronze Age for copper ore mining, through adits on the mountain slopes in the Sinai Peninsula.

Limestone rock as an engineering material was used as a cutting edge for brick caissons in Egypt about 2,000 B.C., to drive a vertical shaft through conglomerate and sand layers until it reached the limestone bedrock. The cutting edge was made of a round limestone block with a vertical hole pierced through its middle (1).

It is interesting to observe that besides serving mining purposes, many tunnels in ancient times were driven also for water supply, public underground passage, erecting of temples, building of tombs, military purposes, and other uses. For example, a road tunnel near Naples, Italy, driven about 36 B.C., was about 1,220 m long, 9.15 m high, and 7.60 m wide.

The first American nonmining tunnel, the Auburn Tunnel, was driven during the years 1818—1821, as part of a canal system in Pennsylvania.

The first American railway tunnel, the Allegheny Portage in Pennsylvania, dates back to 1831—1833.

The Mont Cenis Tunnel, about 11 km long, was the first to pierce the Alps (1857—1870).

The Mont Blanc Tunnel (~11 km long) was driven from 1959—1963.

All these and many other examples of tunneling merely point out the intuitive use of empirical knowledge for utilizing rock for engineering purposes long before science and engineering could supply the logics and theoretical background knowledge. Science moves slowly, creeping on from point to point. The same applies also to rock mechanics.

For the most part, European engineers and geologists in the past 25 years are credited with the development of the principles of rock mechanics, including the methods and techniques of rock testing in situ in connection with applications to a significant number of underground power plants and arch-type dams (Austria, France, Germany, Norway, Sweden, Switzerland, Portugal).

Use of rock mechanics in the United States was mainly limited to mining engineering until the early 1960s. At that time, geotechnical engineers found rock mechanics useful in hard-rock problems. This is somewhat comparable to the application of soil mechanics to problems of earthworks and foundation engineering.

Major organizations in the United States studying rock mechanics are: Society for Rock Mechanics, universities, U.S. Army Corps of Engineers laboratories, Department of Mines, U.S. Department of Transportation, Bureau of Reclamation, National Bureau of Standards, and some private consulting firms.

In 1964, the activities of the American Society for Testing and Materials were broadened to include testing of rocks.

Rock mechanics, first recognized as a distinct discipline only since about 1950, is now advancing in the accuracy, precision, and plausibility of its techniques, and in the rigorousness of its research methods. It can record a considerable progress in theory, experiment, and reliability in testing methods. Especially, progress has been made in determination of stresses in rock; shear strength testing; and reinforcement of rock, or rock bolting.

Reports on research about these topics were presented in three state-of--the-art papers at the 69th Annual Meeting of the ASTM, held June 26 to July 1, 1966, in Atlantic City, New Jersey (2).

The need for the new discipline Rock Mechanics was voiced at the 5th International Conference on Soil Mechanics and Foundation Engineering held in 1961 in Paris, France (16), and was emphasized at the 6th International Conference in 1965 held in Montreal, Canada (17).

No better proof is needed than the fact that the 1st International Congress on Rock Mechanics held in September 1966 at Lisbon, Portugal (18), resulted in three huge volumes of congress papers on rock mechanics dealing with exploration of rock masses; properties of rocks and rock masses; residual stresses in rock masses; comminution; natural and excavated slopes in rocks made in open cuts; underground excavations; deep borings; the behavior of rock masses as structural foundations; and support of massive dams and other engineering structures. Also, brought to the fore was the need for a uniform, standardized, static and dynamic strength test of rocks in situ as well as in the laboratory, and the correlation of such test results.

More than 800 educators, engineers, and engineering geologists from 42 countries attended this historical Congress to review failures of dams and rocks, their causes, and the need for systematically organized knowledge on theoretical, experimental, and applied engineering rock mechanics.

Thus, the First International Congress on Rock Mechanics signified that for the first time in history there was a concerted awareness among engineers (18) and engineering geologists that complex geomechanical problems may respond to quantitative analyses, and that the benefit of such detailed analyses are those of better understanding and consequent better control of natural phenomena encountered in rock engineering.

The U.S. National Research Council's study in 1967 on the value of and need for rock mechanics (14) recommended an acceleration of the total effort in rock mechanics research.

Since 1966, many symposia, conferences, congresses, short courses, and courses on rock mechanics within the course for continued education have been held here and abroad.

International Congresses on Rock Mechanics have been held in Lisbon, Portugal (1966, ref. 18); Belgrad, Yugoslavia (1970, ref. 19); and Denver, Colorado, U.S.A. 1974, (ref. 20).

1-2. Manifestation for the Need of Rock Mechanics

For today's megalopolis, the designing and constructing in rock of underground openings for vehicular and/or aqueous tunnels, underground garages, underpinnings of structures, shelters, storage spaces, underground hydroelectric power plants in connection with requirements for more electrical energy and utilization of "white coal" for that purpose, earth and massive dams, foundations, and other civil engineering structures require the turning over of an enormous amount of earth and rockworks under various and difficult geologic and topographic conditions. Tunnels driven under rivers frequently prove to be more economical than bridges.

As a consequence of the development of our age of specialization, urbanization, industrialization, and conservation of water resources, subsurface excavations in rock are often many times greater now than in the past, and rock slopes associated with highway, railway, causeway, and open-pit engineering are now often several hundred feet high.

Also, it has been said that some phases in progress in engineering, in adding to the store of man's knowledge, have been the results of intensive research on failures of some engineering structures, such as failures of some massive dams founded on rock. Thus, the need for knowledge and experience in engineering rock mechanics is coming more and more to the fore. Large-scale work in rock has been handicapped by lack of understanding of the complex mechanical behavior of the wide variety of geological materials, among rocks. The tremendous scope of engineering projects today dealing with rock as a construction material on a scientific-technical basis requires solutions of engineering problems hardly known before. In the service of safety of life, such engineering activities require civil engineers to design and construct safe and reasonably economical structures. To cope with the unprecedented increase in size and scope of such structures in rock, and to handle strength problems in rocks, today imposes upon the engineer a demand for an extensive, appropriate knowledge not merely of soil mechanics but of rock mechanics as well.

CHAPTER 2

ROCKS

2-1. Definition of Rock

Generally, to the geologist the term "rock" applies to all constituents of the earth's crust. He speaks about "consolidated" rock and "unconsolidated" rock (soil). To the civil engineer, especially the geotechnical engineer, the term "rock" is understood to apply to the hard and solid formations of the earth's crust. The derivatives of rocks, the weathering products, are soils (the unconsolidated sediments). One should be aware that the genesis, viz., formation of geological structures, was not designed by the handbook of some standards. The very heterogeneity of rocks does not lend access to their mathematical analysis and rigorous quantitative study. However, with the future development in rock mechanics which lies ahead of us, hopefully a more exact and quantitative treatment will be possible.

There are several definitions of rock. Some of them are given below.

A rock is a mixture of one or more different minerals. It has no definite chemical composition. Also, it is said that rock is an aggregate of fused or compressed discrete mineral particles. Thus, various minerals are combined in various proportions to form rock.

Broadly speaking, the term "rock" is a general geological term which comprehends any naturally occurring aggregates of minerals or mass of mineral matter, whether or not coherent, constituting an essential part of the earth's crust.

The geologist defines rock as all compositions of the earth's crust onto and into which man builds. Thus, the geologist applies the term "rock" to all constituent parts of the earth's crust.

Rock is also considered to be mineral and organic (soft coal and anthracite, for example) matter that comprehends the solid part of the earth's crust, excluding soil. Terzaghi (25) defines soil as "sediments and other unconsolidated accumulations of solid particles produced by the mechanical or chemical disintegration of rocks." Thus, the distinction between rock and soil is in the degree of consolidation and in the limit of the size of the particles.

A simple concept of rock has been presented by Emery (4), according to whom "Rock is a granular material composed of 'grains and glue.' There is nothing else involved." The "glue" may be ferroginous, calcareous, argillaceous, or siliceous material which cements the grains. Thus, "rock may be described, then, as a granular aelotropic, heterogeneous technical substance which occurs naturally and which is composed of grains of varied polycrystalline or noncrystalline (amorphous) materials which are cemented together either by a *glue* or by a *mechanical bond,* but ultimately by atomic, ionic and molecular bonds within the grains and the glue and at every interface of bonding."

To the engineer who builds on, in, and by means of rock, the term "rock" signifies firm and coherent or consolidated substances that cannot normally be excavated by manual methods alone. To him, rock is a material having manifold properties, like any other material.

In designing in rock, engineers frequently assume rock as a homogeneous and isotropic medium. However, most rocks are not sound; hence they are neither homogeneous nor isotropic.

A *homogeneous* rock material is one in which the physical properties of all of its parts or elements are the same, as opposed to a heterogeneous material.

Heterogeneity (nonhomogeneity or inhomogeneity) is a characteristic of a material, or a force field signifying that the material has properties which vary with position within it (unlike qualities).

An *isotropic* material is one that has the same physical, viz. elastic, properties in every direction at any point. For example, in an elastic medium the velocities of elastic wave propagation are independent of direction.

An *anisotropic* (aleotropic) material is one with certain of its properties varying with direction at any point. For example, many rocks have preferred particle and crystal orientation. Or, because of the state of the stress of the geological formations, sound waves are transmitted through them with different velocities in the horizontal and vertical directions. Hence, such rocks are said to be anisotropic. Therefore they react differently to forces in different directions, depending upon the degree of anisotropy. Thus, the term "anisotropic" is used to mean having different orientations of mechanical properties and fabric.

All these definitions of rock convey that mechanically, rock is a multiple-body system (see Fig. 2-1), and an extremely complex material difficult to work with.

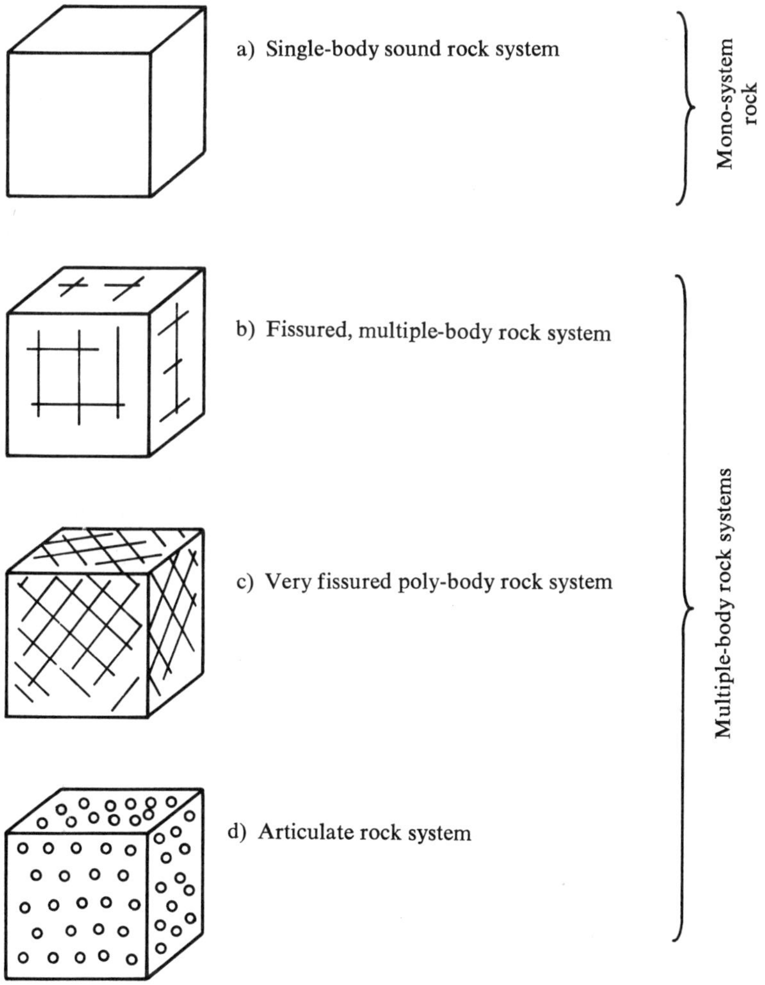

Fig. 2-1
Single and "multiple-body" rock systems

Sometimes rock is defined as a hard, compact material that under stress responds uniformly in deformation and in proportion to the force applied, within certain limits. Such a definition would imply an assumption that rock is an elastic material within these limits. Such an assumption depends to a large degree of course upon the composition and structure of the rock.

A *solid* (compact, fast, firm, rigid, stable, strong) is one which exerts a considerable resistance to separation of its individual particles and/or component parts.

A rock, unlike a steel which can be refined to a consistent internal state before its use, is a naturally occurring material and must be used in its natural state. Certain simplifying assumptions are justified to assist design processes, others are not, and to a large extent the basis for all assumptions lies in the composition and structure of the rock.

As geology has grown more quantitative, more of its parts have become simplified by the reduction of verbal descriptions to numerical measurements, and by the use of mathematics to express relationships among the numbers. However, large areas of geology have stubbornly resisted numerical treatment. The complexity of some problems and the incompleteness of the geologic record limit this approach, even in the age of computers.

Geotechnics

Geotechnics is the civil engineering discipline containing elements of engineering geology fortified with pertinent knowledge from other earth sciences and engineering such as foundation engineering, soil mechanics, rock mechanics, earthworks, hydraulics, hydrology, highway engineering, hydraulic structures engineering (power plants, waterfront structures, tunnels, underground mining openings, for example).

Bedrock

A formation at some depth beneath a mantle of soil is termed bedrock. It is formed from the crust of the earth. It is dry, solid rock exposed to the surface of the earth, or overlain by unconsolidated material.

A *competent rock* is one which is sufficiently strong to transmit a compressive force under given conditions. Rocks which are sufficiently plastic to deform without fracturing are *incompetent*.

Sandstone, quartzite, and igneous rocks are relatively competent under all conditions. Shale and slate are commonly incompetent. Limestone is likely to be competent at low temperature and under moderate pressure. At depth where high pressure and temperature favor recrystallization, rock is likely to be relatively incompetent.

Competent ground is one that does not require support when a tunnel is excavated through it.

An *intact rock* is a rock material which can be sampled and tested in the laboratory, and which is free of the larger-scale structural features such as joints, bedding planes, partings, and shear zones.

Regolith (mantle rock) is loose fragments of rock and soil that act as a cover for bedrock. The soil above the bedrock is called *overburden*.

Rock mass is the in situ rock made up of the rock substance plus the structural discontinuities.

Rock substance is the solid part of the rock mass typically obtained as a drill core.

In rock mechanics, it is necessary to distinguish between rock mass and rock material. Rock material may be defined as the aggregate of mineral particles lying between any structural discontinuity of the rock. Rock mass, in its turn, is the aggregate of regular or irregular blocks of rock material. These blocks are separated by structural features such as bedding planes, joints, fissures, cavities, and other discontinuities.

Rock masses are heterogeneous and usually discontinuous assemblages of rock materials. This means that the scale of an experiment or test with a rock determines to some extent the result of the experiment or test.

2-2. Classification of Rocks

The upper layers of the earth's crust which are accessible to underground and surface rock engineering are in their appearance, stratification, and form of extraordinary variety of petrographical nature and of structural diversity. The structure of rocks, especially, has a marked effect on their strength properties. Hence, there is a need for classification of rocks.

Rocks can be studied most effectively if they are classed according to certain principles into a definite system. A *system* is generally understood to be an ordered grouping of certain elements and/or facts in a field of knowledge according to certain principles.

By its very nature, classification is usually the very first and most important step of activities in the organization of any branch of a scientific and/or engineering discipline. Naturally, therefore, this applies also to rock mechanics in general, as well as to rock classification into certain systems in particular. Rock classification, hence, should be one of the necessary prerequisites in rock engineering.

Rocks may be classified according to several principles. Some of the classification systems are:

1. By origin or genesis,
2. Geological or lithological classification,
3. Engineering classification of intact rock on the basis of rock strength, or
4. A combination of several of these.

1. *By Origin or Genesis*

Based on genesis or mode of origin, rocks are grouped into three broad groups:

Igneous rocks,
Sedimentary rocks, and
Metamorphic rocks.

Igneous rocks form when hot molten silicate material from within the earth's crust solidifies.

Sedimentary rocks form from deposition and accumulation of sediments of other rocks, plant remains, and animal remains by wind, or water at the earth's surface, and their later solidification into rock.

Metamorphic rocks form when already existing rocks undergo changes by recrystallization in the solid state at high pressure, temperature, and chemical action at some time in their geological history.

2. *Geological or Lithological Classification*

Lithology is the study of the physical character of rocks. The lithology of a rock pertains to its mineralogical composition and texture together with a descriptive term from some accepted rock classification system, i.e., granite, limestone or, say mica schist. Thus, the terms and classification are geological.

Although in several ways the lithological rock classification has some advantages, for engineering purposes the geologic term alone is insufficient for engineering classification of rocks.

3. *Engineering Classification of Intact Rock*

This classification system is here presented according to DEERE's (3) lectures given at the ASCE Rock Mechanics Seminar April 9—10 and May 7—8, 1968.

The classification system pertains to intact rock and to rock in situ. Intact-rock classification presupposes rock-testing. The classification is based on two important engineering properties of rock, namely: the uniaxial compressive strength, and the tangent modulus of elasticity (taken at a stress level equal to one-half the ultimate compressive strength of the rock). Table 2-1 shows the engineering classification of intact rock on the basis of uniaxial compressive strength after DEERE and MILLER (3).

TABLE 2-1

Engineering Classification of Intact Rock
(after DEERE and MILLER, Ref. 3)

1. On the basis of strength

Class	Description	Uniaxial Compressive Strength psi	Uniaxial Compressive Strength MN/m^2	Rock Material
A	Very high strength	> 32,000	~ 220	Quartzite; diabase; dense basalts
B	High strength	16,000–32,000	~ 110 to ~ 220	Majority of igneous rocks; strong metamorphic rocks; weakly cemented sandstones; hard shales; majority of limestones; dolomites
C	Medium strength	8,000–16,000	~ 55 to ~ 110	Many shales; porous sandstones and limestones; schistose varieties of metamorphic rocks
D	Low strength	4,000–8,000	~ 28 to ~ 55	Porous low-density rocks; friable sandstone; tuff;
E	Very low strength	< 4,000	< 28	clay shales; weathered and chemically altered rocks of any lithology

In this table, the rock-strength classes follow a geometric progression.

Table 2-2 shows the engineering classification of intact rock on the basis of modulus ratio $M_R = E_{t\,50}/\sigma_{ult}$ after DEERE and MILLER (3), where $E_{t\,50}$ = tangent modulus at 50% ultimate compressive strength of the rock, and σ_{ult} = uniaxial ultimate compressive strength.

Based on this classification, rocks are classified both by strength and modulus ratio as AM, BL, BH, and CM, for example. The classification is also sensitive relative to mineralogy, texture, and direction of anisotropy.

TABLE 2-2

Engineering Classification of Intact Rock
(after DEERE and MILLER, Ref. 3)

2. On the basis of modulus ratio

Class	Description	Modulus ratio, E_{t50}/σ_{ult}
H	High	> 500
M	Average (medium)	200–500
L	Low	< 200

2-3. Rock as a Construction Material

To a civil engineer who lays foundations for structures on rocks and builds in rock, rock is an engineering material. As naturally occurring material, rocks have many uses and applications. For example, rock is used:

1. For laying structural foundations to support structures;
2. For constructing vehicular and equeous tunnels, underground power plants, and other underground openings and cavities;
3. For building moles or breakwaters;
4. For protective blanketing of earth dams and other earthworks against erosion by water in the form of riprap;
5. For ballast on railroads in the form of crushed rock;
6. As a base and subbase courses for roads and airfield runways in the form of crushed rock;
7. As an aggregate for concrete (crushed rock);
8. As a dimension stone;
9. As facing stone for buildings, bridges, and hydraulic structures to protect these structures from weathering;
10. For making artifical sand from rock in regions poor in sand but rich in suitable rocks.

These are only a few uses for rock as an engineering material. There may be many more.

Without natural stones, the building of highways, railroads, airfields, bridges, harbors, industrial structures, and residential and cultural edifices, the regulation of rivers and the entire field of concrete engineering would be impossible.

As engineering materials, rocks must be indentified, described, tested, and classified.

To utilize rock as a construction material in design and for building of safe and economical structures, its physical and mechanical properties must be known and stress conditions in rock *en masse* must be understood.

Rock as a construction material, or a rock mass, frequently contains large quantities of water and various tectonic discontinuities, and may be subjected to high *confining pressures*—all of which may radically alter its properties, which are normally based on small-scale laboratory test data.

Among the important questions which the geotechnical engineer has to consider in his work are:

1. The structure of rocks in relation to tunneling operations;
2. Dam, reservoir and building foundations;
3. Landslides (stability of slopes);
4. The nature or character of the common rocks in their use for building stone and road material; and
5. The geological conditions affecting and controlling underground water supplies.

CHAPTER 3

ROCK MECHANICS

3-1. Definition of Rock Mechanics

Let us now fix the idea of rock mechanics and its position.

Rock mechanics is principally characterized by the fact that rock is not a continuum but a regulated discontinuum, also called articulate, grain-structure mechanics.

In a large, solid, sound (healthy) mass, rock may be considered to be a continuum. Rock mechanics is the mechanics of the discontinuum.

Whereas soil mechanics works in plane, rock mechanics works both in plane and in space.

According to the definition put forward by the Committee on Rock Mechanics of the Geological Society of America in 1964, viz., of the National Academy of Sciences Committee on Rock Mechanics in 1966 (13): "Rock Mechanics is the theoretical and applied behavior of rock; it is that branch of mechanics concerned with the response of rock to the force fields of its environment."

According to JUDD (8), every word in this definition was carefully evaluated to ensure a minimum of misunderstanding. No attempt was made to qualify the term "rock". However, by the logical contents of the matter, the subject rock mechanics pertains not merely to mechanics, but to the geologic materials as well that exist in nature or have been obtained from a geological environment. Although intimately concerned with the same natural, geological material, the new engineering discipline rock mechanics does not lie in the province of seeking explanation about the origin of geological structures as the scientific discipline geology does. Also, rock mechanics is an engineering field of its own, distinct form engineering geology. Engineering rock mechanics deals with rock as an engineering material. It also deals primarily with those changes in mechanical behavior in rock such as stress, strain, and motion of rocks brought about by engineering activities. Among other things, the subject of rock mechanics concerns itself with the design and stability of underground structures in rocks.

As an engineering discipline, rock mechanics is not an abstract study in mathematics and/or mechanics; it is a high-level applied engineering science. It can be successful only when connected with the real geological situation when designing and constructing foundations on rock and substructures in rock.

Thus, to the engineer, the subject of rock mechanics involves the analysis of static and/or dynamic loads or forces as applied to the rock, the analysis of the internal effects in terms of either stress, strain, or stored energy; and the analysis of the consequences of these internal effects, i.e. fracture, flow, or simply deformation of the rock. One, therefore, may say that rock mechanics is the youngest branch of the discipline strength of materials.

As in the field of soil mechanics and other mechanics, rock mechanics makes an extensive use of the theory of elasticity and the theory of plasticity, and studies rocks and structure-rock systems experimentally.

One sees that rock mechanics is a maturing engineering discipline, emphasizing its breath in both theoretical and applied areas. Rock mechanics is important within the scope of progress in the geotechnical sciences.

3-2. Some Features of Rock Mechanics

1. In a large, solid, sound rock mass, rock may be considered to be a continuum.

2. However, in its natural, geologic environment, rock is principally characterized by the fact that rock is not a continuum but a regulated discontinuum because of joints, fissures, shistosity, cracks, cavities, and other possible discontinuities. In this respect, it may be said that under certain conditions, rock mechanics is the mechanics of the discontinuum or, in other words, the mechanics of the structure of the rock.

3. Mechanically, rock is a "multiple-body" system.

4. Soil mechanics analyses work mostly in plane, whereas rock mechanics works both in plane and in space.

5. Rock mechanics has developed independently of soil mechanics. Of course, there is a certain overlap between these two engineering sciences.

A civil engineer in the fields of geotechnics and construction must know the material with which, upon which, and in which he works.

Rock, as one of the most commonly used construction materials, is very complex and difficult to describe and define. It is a material about which we know very little. To avoid structural failures associated with rock, it is an absolute necessity to understand it and to control it, especially in its natural environment under the influence of load, water, and temperature.

3-3. Rock Mechanics Problems

Some of the problems in rock mechanics associated with man's activity on and in rock, and the relationships that exist between a foundation on or an underground opening in rock, may be listed as follows:

How will rock react when put to man's use?
What is the bearing capacity of a rock on its surface and at various dephts to carry various loads?
What is the shear strength of rock?
How will rock perform under dynamic loads?
What is the effect of earthquakes on a rock-foundation system?
What is the modulus of elasticity of rocks?
What is the POISSON's ratio of rocks?
What are the effects of rock defects (jointing, bedding planes, schistosity, fissures, cavities, and other discontinuities) on its strength properties?
Which laboratory test method of rocks most nearly provides the actual in-situ properties of a foundation-supporting rock *en masse?*
How to account for joints and faults in design in rock?
How to cope with time-dependent deformations (creep) in rock?
What are the laws of plastic flow of rocks?
What is the effect of anisotropy of rock on stress distribution in rock?
How to correlate the results of rock strength tests made in situ with those made in the laboratory on prepared rock specimens?
What test method would most readily provide the actual in-situ conditions and properties of a rock?
What are the mechanisms of failure of rocks?
Can the state of stress in rock be accurately estimated, or can it even be measured?
What are the rock-slope design factors?
Will roof bolting in an underground opening in rock provide a satisfactorily safe permanent installation?

Also, there is the problem as to the factor of safety to use in design in rock.

The engineer is also interested in the water regimen in rock, in the mutual interaction between rock and foundation; suitability of rock as a construction material; effect of weathering on soil; and grouting of fissured rock to improve its performance relative to load and uplift pressures.

From these questions, one can see that "rock mechanics versus foundation" problems are attached to the rock, and that they should be studied as problems of deformation, strength, and stability.

The scope of the contents of rock mechanics, of course, overlaps several other disciplines of engineering and science (soil mechanics, engineering geology, geophysics, for example) which, in the past were not always considered to be too closely related.

The rock mechanics problems could be solved reasonably accurately:

1. If the stresses in a loaded rock were possible to indicate even approximately (for example, analytically), and

2. If for the induced stresses and stress distribution in rock to be analyzed, it were possible to indicate, even approximately, the corresponding factor of safety against rupture or dangerous deformations.

Although the technology strives for quantifying knowledge of the *en masse* properties of rock material and a plausilbe comprehension of stress distribution within it, unfortunately, even today, no exhaustive solution to these problems can be given as yet, mainly because of the inhomogeneity and anisotropy of the rock material and, in the words of Stini (23), because "Nature is different everywhere, and she does not follow textbooks." In this respect, rock mechanics has not reached a state of completeness as yet. These problems cannot be solved plausibly by any formulas as yet, but require some idealization of the rock material and the consideration of the effects of many environmental factors, without the possibility of an exact proof of stresses as a soft cushion of our conscience.

Besides, the problem of stress distribution in rock is a hint that the difficult problems of construction in rock cannot be simply burdened upon the geologist. Instead, the rock engineer should have the information on rock qualities as observed by the geologist, to evaluate numerically in his conclusions.

3-4. Objectives of Rock Mechanics

Some of the objectives of engineering rock mechanics are:

1. To perform engineering rock surveys;

2. To develop rational rock sampling, identification and classification methods;

3. To develop suitable rock testing devices and standard testing methods for compressive strength and shear strength of rock materials;

4. To collect and classify information on rocks and their physical properties in the light of fundamental knowledge of rock mechanics, foundation engineering and hydraulic structures engineering;

5. On the basis of test results, to study the physical, mechanical (static and dynamic), elastic, inelastic, plastic and rheological properties of rocks, and their mode of failure under static and dynamic loading;

6. To study rock performance under thermal conditions and water regimen;

7. To deal with statics and dynamics of structures in rock;

8. To develop methods of in-situ measurement of static and dynamic deformation properties of rock and residual stresses in rock under various environmental conditions of weathering, leaching, seismics, and tectonics;

9. To perform research on the mechanisms of failure of rocks;

10. To organize research on rock reinforcement and for in-situ stress measurements;

11. To replace by scientific methods the empirical ones of design used in rock engineering in the past, thus contributing to the advancement of the rock mechanics discipline;

12. To stimulate and disseminate knowledge on rocks and rock mechanics;

13. To apply the knowledge of rock mechanics for the solution of practical engineering problems;

14. To study performance of natural intact rock *(en masse)* under load and environmental conditions; and

15. To deal with statics of structures in rock; stability of rocks is of utmost importance from the viewpoint of both safety and economy.

Thus, by and large, the objective of engineering rock mechanics is its application to the solution of geotechnical problems.

International and national committees are busy planning *specifications* for *classification* of rocks, and for various laboratory and in-situ techniques which can be used to develop a better awareness of the action of rock under various environments.

3-5. Value of Rock Mechanics

Just as the value of soil mechanics is indisputable in soil engineering, so is the value of rock mechanics in rock engineering. Rock mechanics is one of the most useful aids to foundation engineers in their everyday professional activities.

3-6. Theoretical Basis of Rock Mechanics

Just as every other engineering discipline has a theoretical basis, so has rock mechanics. As is so often the case in a relatively new discipline, correct solutions to problems may require the engineer and researcher to borrow some general knowledge and theories from other disciplines and to adjust them to suit their particular field in forming a theoretical basis. The theoretical basis is needed for the ultimate understanding of the strength of the rock and its performance in situ. Inasmuch as every physical law has been derived from experiment, experiments with rocks play an important part in rock mechanics. Also, experiments with rocks serve for checking certain aspects of adopted theories from other disciplines. However, according to Sir ROGER BACON, experimental science does not receive knowledge from any higher sciences: experimental science is the mistress; all other sciences are her servants.

General Notes

Many rock mechanics and rock engineering problems of practical importance involve both elasticity and plasticity of rocks, especially when concerned with strength and safety of an underground opening in rock, or with the stability of rock as a material supporting structural foundations.

Rock mechanics, like any other engineering discipline, has a theoretical basis. In rock mechanics, frequently the theory of elasticity and the theory of plasticity are used as a part of that basis. Therefore, the acquaintance with some basic concepts of these theories is desirable.

Let us first recall that elasticity and plasticity are important universal properties of crystalline and amorphous solids, among rocks.

Frequently, a rock mechanics problem on hand calls for quantitative studies: first with all parts considered to be within the range of elasticity, and thereafter with some parts admitted to be in the range of plasticity (refer to Chapters 7, 8, and 9 in this book).

Both the theory of elasticity and the theory of plasticity deal with stresses and deformations in solid materials.

In their present state, the theory of elasticity and the theory of plasticity arer concerned mostly with problems of loads and equilibrium within the realm of statics.

Theory of Elasticity

The essential store of all knowledge of structural theory, design, and construction practice is the mathematical theory of elasticity.

A review of the development of the theory of elasticity shows that during a period of over 250 years, the theory of elasticity had gradually developed into an exact discipline as a part of mechanics. Today, mechanics is the solid foundation for the design of engineering structures.

In the mathematical theory of elasticity, an ideal, homogeneous, elastic, isotropic medium (viz., material) is postulated. The theory of elasticity is based on HOOKE's law of linear proportionality E between stress σ and strain ε, i.e.,

$$\sigma = E \cdot \varepsilon \tag{3-1}$$

Ut tensio sic vis

Freely translated, HOOKE's law reads: As the stress, so the strain.

The coefficient of proportionality E is called the modulus of elasticity of the material.

The concepts of stress and strain are fundamental in the theory as well in the design of engineering structures.

It is also assumed in this theory that the deformation, viz., strains of a loaded material, are small quantities. Hence, they can be neglected in setting up equations of equilibrium.

It is further assumed in this theory that in the externally loaded elastic material, all strains are instantaneously and totally recoverable upon removal of all external load. A material is perfectly elastic if it recovers completely upon removal of all external stresses.

In rock mechanics, the object of a problem relating to elasticity is usually to ascertain the stress distribution in an elastic material. In some other cases, strains must be found at any point brought about by given body forces and given conditions at the boundary of that body. In such elasticity problems, one is dealing with infinitesimal strains and displacements.

According to NÁDAI (11), these assumptions are supported by actual measurements of the observed strains or displacements within the elastic range in metals, rocks, and other materials, to a sufficient degree for most practical applications.

But rock is not exactly an ideal, elastic material. An elastic medium is an idealization of actual material properties, for obviously, deformation of crystalline materials under the action of external forces is the result of distortion of the crystal lattice in which the atoms, ions, and/or molecules are rearranged, involving structural adjustments inside the rock material, which adjustments will be a finite process. Hence, there will be less than a total recovery of deformations after the removal of the externally applied load from the material.

Also, the concept of isotropy is applied in rock mechanics as a simplifying assumption used in continuum mechanics. Most crystals and rocks are not isotropic.

The main objections against the use of the theory of elasticity in rock mechanics are

1. Because of the presence of discontinuities and various rock defects in the rock in situ, thus affecting the mechanical properties of the rock (rock has a grained structure consisting of minerals and crystals of several kinds with many different orientation, and with irregular boundaries between the grains):
2. Because of the ever present internal (residual) stresses in the rock mass; and
3. Because of the dependence on time of the deformations (rheological aspects).

Unfortunately, for rocks, frequently the established functions to use are so complex and involved, that in attempting to seek a reasonable and acceptable solution one must resort to idealizations and simplifications; or to the method of trial and adjustment; or to graphoanalytical methods; or in-situ and laboratory testing of rocks; or experimental model testing of prototype structure.

Nevertheless, to a certain extent, the theory of elasticity and the theory of plasticity are used in the design of structures in rock. Here, simplified rock properties are assumed in order to assist analytical calculations based on the theory of elasticity and/or plasticity. According to FARMER (5), sometimes such assumptions are successful and sometimes not. It is therefore essential that the limits of applicability of the theory of elasticity to rock be clearly defined.

If the laws of HOOKE's linear elasticity, and those of ideal plasticity, are compared with the typical experimental features, one can determine the applicability and limitations of the mathematical formulations for rock. By observing near the origin of the stress-strain diagrams for rocks, one notices that the linear stress-strain relationship is generally good in nearly all cases.

Another point to make here in favor of applicability of the theory of elasticity to rock mechanics is as follows: the stress-strain relationships as derived on the basis of the theory of elasticity alone, are suitable for the theoretical verification of the development of the stress-free body around underground openings, and the stresses should be carried entirely by the rock surrounding the opening.

For rock engineering purposes, geological and rock mechanics data, results obtained from in-situ testing of rocks, and analyses of strength properties of rocks buttressed by model tests, coupled with experience and judicial judgement, naturally form the basis for the design and construction in rock.

The reader should not infer that the above made assumptions are a rigid framework limiting the scope and applicability of rock mechanics. Unless simplifications represent the real conditions in rock with sufficient accuracy, the physical laws governing the nature of the performance of the rock-structure-load system should not be arbitrarily twisted in order to obtain a simplified working system that could be treated easily mathematically, or be "in good agreement", or would render in advance a preconceived result.

Theory of Plasticity

The object of the theory of plasticity is to study, mathematically, stresses and displacements in plastically deforming materials. The basic feature of the theory of plasticity as compared with the theory of elasticity is the non-linear relationship between stress and strain.

In contradistinction to the theory of elasticity, in the theory of plasticity one considers bodies in which, under the action of large enough stresses, there set in permanent deformations which remain after the removal of the externally applied load. Actually, all materials possess a certain degree of plasticity, i.e., property of preserving a certain part of a deformation after relief of stress.

The theory of plasticity as applied to an ideal rock assumes that on a certain plane in rock there exists a state of failure or rupture at any point located on such a plane. Such a theory has been put forward by Hencky (7), Nádai (11), Prandtl (15), and others.

Prandtl contributed to the theory of plastic equilibrium of a horizontally bounded hemispatial medium subjected to a vertical punching pressure (15).

Other authorities on plasticity and rupture are Saint Venant (22), von Mises (10), Kármán (19), Tresca (26), and others.

Generally, the theory of plasticity is concerned with all deformations that are independent of time. The explanation of plastic flow in solid materials is the permanent displacement in the relative positions of the elements or atoms in the crystal lattice.

The theory of plasticity is confined to equilibrium within the stressed body. Thus, the theory can determine only the state of stress, both in the plastic and elastic zones, after all deformations have ceased to change with time.

Plastic deformation is physically anisotropic. Any initial isotropy which may have been present is usually destroyed by plastic deformation.

Within the range of plasticity, the relationships between stress and strain are more complex than within the elastic range. The strains are functions not only of the corresponding acting stress and temperature, but also of the past history of stresses, such as those which have been brought about by tectonic activities in the earth's crust.

The study of the strength properties of rocks is based on the conclusions of the theory of plasticity, because fracture, as a rule, precedes plastic deformations.

Materials with polycrystalline grain structure, such as minerals and brittle rocks, under certain conditions may be brought into the plastic state, in which permanent deformations may occur without fracture of the material. Solid materials in this condition are said to yield, or to deform plastically.

In describing the plastic states of stress in rock, extensive use is made of the slip surfaces (Chapter 8).

The theory of plasticity has yielded powerful general concepts that have become tools of enormous importance in the rational design of a wide range of structures, among them structures in rock. In civil engineering, the theory of plasticity is applied to soil mechanics and foundation engineering, and to rock mechanics as well. Plasticity is also of great interest to the mining industry, as well as to geology and to geophysics.

Difference between the Engineering Discipline of Strength of Materials and Rock Mechanics

Functionally, the ground (soil and/or rock) belongs to structure, and vice versa. The ground and the structure are a mutually interacting integral system called the rock-structure-load system. This system includes all the effects of discontinuities, viz., rock weaknesses or defects. In rock mechanics, therefore, consideration of relevant important engineering concepts should not be omitted from the presentation that follows. However, corresponding to the sense of content and intent of this book on rock mechanics, the engineering geology concepts are subordinated to some extent under the aegis of engineering mechanics within the scope of rock mechanics.

Particularly in the phase of mechanics within the discipline of rock mechanics, elements from mechanics of materials, theory of elasticity, theory of plasticity, theory of mechanics of fracture, rheology, and other branches of scientific-technical knowledge are involved.

In the wake of the above, naturally some questions arise, such as:

What is rock mechanics?
What is the difference between strength of materials and rock mechanics?

The definition of rock mechanics was given earlier in Chapter 3 of this book. However, popularly speaking, rock mechanics may be regarded as the strength of materials of rocks. The strength of the rock is its ability to resist externally applied loads.

In the discipline of strength of materials, called also the mechanics of materials, a load is applied to a *stress-free* material, and the so imparted stresses and strains in it are analyzed. Here, the external loading and the geometric dimensions of the loaded structural element are known.

In rock mechanics (refer to Section 7), the rock in situ is initially subjected to an *initial* or *primary stress field*. Upon introducing an underground opening in the rock mass, a new stress condition in rock is brought about, namely: stress is released locally (stress reduction or stress relaxation), resulting in a corresponding unloading deformation.

One notices that the basic difference between strength of materials and rock mechanics is the changed stress conditions and deformations in the body of the material brought about by an excavated underground opening. Thus, rock mechanics has for its purpose to analyze newly induced stresses and strains in the rock massif brought about by an underground opening (see Fig. 7-3).

The geometric dimensions of the region or zone of disturbance in the rock around the cavity that takes up the unloading deformation must first be determined (Fig. 7-3), which is difficult to do. It can thus be reasoned that the load and the load-bearing structural element cannot be precisely defined. Also, one bears in mind that rock as a construction material is not a homogeneous material manufactured artificially according to certain standards. On the contrary, rock is a natural, nonhomogeneous material. Besides, of special significance is the fact that the rock deformations, brought about because of the changed stress conditions, are seldom of elastic nature. In most cases they are of plastic nature, or at least partly elastic and partly plastic. Thus the discipline of practical engineering rock mechanics in situ pertains actually to the plastic zones in the rock. Because of their complexity, the plastic zones are difficult to comprehend analytically. Therefore, approximation or simplification of the more exact theory is made, so that the practicing engineer can utilize the predictions and results of the mathematical theory of plasticity.

Because the mechanical phenomenon of pressure on tunnels may have several different causes, there exist many methods of calculations of stress conditions around various forms of tunnels (6, 21, 24). Besides, the striving after a universal, omnipotential formula for computing the magnitude of the pressure on underground openings and lined and walled tunnels cannot conform exactly with the reality.

3-7. Rock Engineering

The discipline of engineering rock mechanics is the scientific basis for rock engineering. Rock engineering is the practical, technical use and engineering application of rock mechanics in the design of engineering structures such as various underground openings as set forth: power plants, storage space, protective shelters, fortifications, shafts, adits, mines, vehicular and aqueous tunnels; in the utilization of rock as a support of structural foundations; or in the design of roads, railroads, canals, concrete dams on rock, quarries, open excavations, design of rock slopes in cuts, anchorage of rocks.

Just as soil engineering was influenced decisively by new concepts in the field of soil mechanics, so it is with rock engineering: through research, experience, and accumulation of man's knowledge of the physical and strength properties of rocks, a new engineering discipline of rock mechanics and rock engineering is coming to the fore.

Although a relatively new discipline still in need of development, today rock mechanics and rock engineering are recognized as distinct geotechnical engineering disciplines in their own rights, just as soil mechanics and foundation engineering.

3-8. Ice Mechanics

The place of ice mechanics is between soil mechanics and rock mechanics. Geologically, ice is a rock (the solid phase of water). As a rock, ice has been recognized by the First Congress of the International Society of Rock Mechanics held September 25 to October 1, 1966, at Lisbon, Portugal.

REFERENCES

1. Anon., *Engineering News Record,* December 7, 1933, p. 675.
2. ASTM, *Proceedings of the 69th Annual Meeting of the ASTM,* held June 26 to July 1, 1966 in Atlantic City, New Jersey. Philadelphia, Pa., 1966.
3. Deere, D. U., and R. P. Miller, 1966, "Engineering Classification and Index Properties for Intact Rock," *Technical Report No. AFWL-TR-65-116,* Air Force Weapons Laboratory, Kirtland Air Force Base, New Mexico.
4. Emery, C. L., 1966: "The Strain in Rocks in Relation to Highway Design." HRB, Record No. 135 on *Rock Mechanics,* National Research Council, Washington, D. C., pp. 1–9.
5. Farmer, I. W., 1968, *Engineering Properties of Rocks,* London: E. and F. N. Spon Ltd., p. 40.
6. Fenner, R., 1938, "Untersuchungen zur Erkenntnis des Gebirgsdrucks." *Glückauf,* No. 32, pp. 681–695.
   No. 33, pp. 705–715.
   Study of Ground Pressures. Translated from the German by W. Kochan and B. Korun, Ottawa, National Research Council of Canada, Technical Translation TT–515, 1955.
7. Hencky, H., 1923, "Über einige statisch bestimmte Fälle des Gleichgewichts in plastischen Körpern." *Zeitschrift für angewandte Mathematik und Mechanik,* No. 3, pp. 241–351.
8. Judd, W. R., ed., 1964: *State of Stress in the Earth's Crust.* Proceedings of the International Conference, held June 13–14, 1963 at Santa Monica, California. New York, N. Y.: American Elsevier Publishing Company, Inc.
9. Kárman, Th., 1911, "Festigkeitsversuche unter allseitigem Druck." *Zeitschrift des Vereins deutscher Ingenieure,* Vol. 55, pp. 1749–57.
10. Mises, R., 1913, "Mechanik der festen Körper im plastisch-deformablen Zustand," Göttingen. *Nachrichten der Gesellschaft der Wissenschaften zu Göttingen.* Mathematisch physikalische Klasse, pp. 582–592.
11. Nádai, A., 1950, *Theory of Flow and Fracture of Solids,* New York: McGraw-Hill Book Company, Inc., vol. 1, 2nd edition.
12. Nasiatka, Th. M., 1968, *Tunneling Technology,* Washington, D.C.: U.S. Department of the Interior, Bureau of Mines, U.S. Government Printing Office. Information Circular 8375.
13. National Academy of Sciences: *"Rock Mechanics Research,"* NAS-National Research Council, Publication 1466, Washington, D.C., 1966.
14. National Academy of Sciences-National Academy of Engineering, National Research Council: *News Report XVII,* No. 2, February, 1967, p. 4.
15. Prandtl, L., 1921, "Über die Eindringungsfestigkeit (Härte) plastischer Baustoffe und die Festigkeit von Schneiden," *Zeitschrift für angewandte Mathematik und Mechanik,* vol. 1, No. 1, February 1921, pp. 15–20.
16. *Proceedings of the 5th International Conference on Soil Mechanics and Foundation Engineering,* held July 17 to 22, 1961 in Paris, France. Published by Dunod, Paris, 1961.

17. *Proceedings of the 6th International Conference on Soil Mechanics and Foundation Engineering,* held September 8 to 15, 1965 at Montreal, Canada. Published by the University of Toronto Press.

18. *Proceedings of the (First) Congress of the International Society of Rock Mechanics, 1966,* held September 25 to October 1, 1966 at Lisbon, Portugal. Published by the Laboratório Nacional de Engenharia Civil, Lisbon, Portugal (3 volumes: vols. 1 and 2–1966; vol. 3–1967).

19. *Proceedings of the Second Congress of the International Society of Rock Mechanics, 1970,* held September 21–26, 1970, at Belgrade, Yugoslavia. Published by the Institute for Development of Water Resources. Jugoslav CERNI in 1971 at Belgrade.

20. *Proceedings of the Third Congress of the International Society for Rock Mechanics,* 1974, held Sept. 1–7, 1974, at Denver, Colorado. Published by the National Academy of Sciences, Washington, D.C.

21. PROCTOR, R. V., and T. L. WHITE, 1946. *Rock Tunneling with Steel Supports,* with an Introduction to Tunnel Geology by KARL TERZAGHI. Youngstown, Ohio: The Commercial Shearing and Stamping Co.

22. SAINT-VENANT, B. DE, 1970, "Sur l'établissiment des equations des mouvements intérieurs opérés dans les corps solides ductiles au delà des limites où l'élasticité pourrait les ramener à leur premier état." *Comptes Rendus,* Paris, vol. 70, 1870, pp. 473–480.

23. STINI, J., 1952, "Der Gebirgsdruck und seine Berechnung," *Geologie und Bauwesen,* No. 3, 1952, pp. 165–200.

24. SZÉCHY, K., 1966, *The Art of Tunneling.* Akadémiai Kiadó, Budapest.

25. TERZAGHI, K., 1943, *Theoretical Soil Mechanics.* New York, N.Y.: John Wiley and Sons, p. 1.

26. TRESCA, H., 1864, "Mémoire sur l'écoulement des corps solides soumis à des fortes pressions." *Comptes Rendus,* Académie des Sciences, Paris, July–December, vol. 59, 1854, pp. 754–758; vol. 64, 1867, p. 809.

OTHER RELATED REFERENCES

ASCE – *Symposium on Underground Rock Chambers,* held January 13–14, 1971, at Phoenix, Ariz.

ASCE – *New Horizons in Rock Mechanics,* 14th Symposium on Rock Mechanics, 1972 (785 pages).

ASCE – *Applications of Rock Mechanics,* 15th Symposium on Rock Mechanics, 1973 (650 pages).

ASCE – *Lessons from Dam Incidents,* 1973 USCOLD Report (387 pages).

ASCE – *Foundations for Dams,* 1974, Engineering Foundation Conference (472 pages).

ASCE – *Design Methods in Rock Mechanics,* 16th Symposium on Rock Mechanics, 1975 (422 pages).

ASCE – *Rock Engineering for Foundations and Slopes,* 2 volumes, 1976 Conference (879 pages).

ASCE – *Numerical Methods in Geotechnics,* 3 volumes, 1976 Conference (1,517 pages).

ASCE – Committee on Continuing Education of the Foundations and Soil Mechanics Group, 1976, *Rock Excavation* Seminar. New York, N.Y.: ASCE Metropolitan Section, October 14, 21 and 28, 1976.

ASCE – *Rock Engineering for Foundations and Slopes*. New York, N.Y.: Proceedings of the ASCE Specialty Conference held August 15–18, 1976, at University of Colorado, Boulder, Colo., Vol. I, 1976, and Vol. 2, 1977.

ASTM – *1977 Annual Book of ASTM Standards,* Part 19: Natural Building Stones; Soil and Rock; Peats, Mosses, and Humus. Philadelphia, Pa.: ASTM, 1977.

ATTEWELL, P. B., and I. W. FARMER, 1976, *Principles of Engineering Geology*. London: Chapman and Hall and John Wiley and Sons, Inc.

BAKER, R. F. et al., 1972, *The Use of Underground Space to Achieve National Goals* (A Report of a Program for Improving the Effectiveness of Underground Construction), ASCE, in cooperation with American Institute of Mining, and Metallurgical and Petroleum Engineers, sponsored by the National Science Foundation, Washington, D. C., and the Engineering Foundation, New York, N.Y., December 31, 1972.

BLYTH, F. G. H. and M. H. DE FREITAS, 1974, *A Geology for Engineers*. London: Edward Arnold Ltd.

BORG, S. F., 1962, *Fundamentals of Engineering Elasticity*. New York, N.Y.: D. Van Nostrand Company, Inc.

Channel Tunnel Study Group, 1960, "Report of the Channel Tunnel Study Group," Channel Tunnel Company, London.

COATES, D. F., 1970, *Rock Mechanics Principles*. Ottawa: Information Canada, Mines Branch Monograph 874.

Committee on Rock Mechanics, 1966, *Rock Mechanics Research*. New York, N. Y.-Washington, D.C.: National Academy of Sciences-National Research Council, Publication No. 1466.

DREYER, W., 1972, *The Science of Rock Mechanics*. Clausthal, Germany: Trans Tech Publications.

DUNCAN, N., 1969, *Engineering Geology and Rock Mechanics,* 2 volumes. London: Leonard Hill.

FARMER, J. W., 1968, *Engineering Properties of Rocks*. London: E. & F. N. Spon Ltd.

FELD, J., 1966, *Rock as an Engineering Material*. Evanston, III.: Soiltest, Inc. (32 pages).

The Geological Society of America (1950/55 and other years), *Application of Geology to Engineering Practice*. New York, N.Y.: The Geological Society of America.

HEYMAN, J., and F. A. LECKIE (editors), 1968, *Engineering Elasticity*. Cambridge University Press (706 pages).

Highway Research Board, 1958, Special Report 29, edited by E. B. ECKEL, *Landslides and Engineering Practice*. Washington, D.C.: National Academy of Sciences-National Research Council, Publication 544.

HILL, R., 1950, *The Mathematical Theory of Plasticity*. Oxford: Clarendon Press.

HIX, C. F., Jr. and R. P. ALLEY, 1958, *Physical Laws and Effects*. New York, N.Y.: John Wiley and Sons, Inc., p. 64.

JAEGER, C., 1961, "Rock Mechanics and Hydro-Power Engineering", *Water Power,* Vol. 13, Nos. 9 and 10, September-October, 1961.

JAEGER, C., 1972, *Rock Mechanics and Engineering*. Cambridge University Press.

JAEGER, J. C., 1962, *Elasticity, Fracture and Flow* (2nd ed.). London: Mathuen & Company Ltd.

JAEGER, J. C., and N. G. W. COOK, 1971, *Fundamentals of Rock Mechanics*. London: Chapman and Hall Ltd., pp. 85–99.

JOHN, K. W., 1962, "An Approach to Rock Mechanics", *Proceedings ASCE,* Soil Mechanics and Foundations Division, Vol. 88, No. SM 4, Part I, pp. 1–30.

JOHNSON, A. M., 1970, *Physical Processes in Geology.* San Francisco, Calif.: Freeman, Cooper and Company.

JUMIKIS, A. R., 1969, *Theoretical Soil Mechanics.* New York, N.Y.: Van Nostrand Reinhold Company.

JUMIKIS, A. R., 1971, *Foundation Engineering.* Scranton, Pa.: Intext Educational Publishers.

JUMIKIS, A. R., 1973, *Settlement Tables for Centrically Loaded, Rigid Circular Footings on Multilayered Soil Systems.* New Brunswick, N. J.: Rutgers University, College of Engineering, Bureau of Engineering Research, Engineering Research Bulletin 54, pp. 20–22.

KRYNINE, D. P., and W. R. JUDD, 1957, *Principles of Engineering Geology and Geotechnics.* New York, N.Y.: McGraw-Hill Book Company, Inc.

LEGGET, R. F., 1962, *Geology and Engineering* (2nd ed.). New York, N.Y.: McGraw-Hill Book Company, Inc.

LOVE, A. E. H., 1944, *A Treatise on the Mathematical Theory of Elasticity* (4th ed.). New York, N.Y.: Dover Publications.

MOOS, A., and F. DE QUERVAIN, 1948, *Technische Gesteinskunde.* Basel: Birkhäuser, Mineralogisch-Geotechnische Reihe I, Vol. 15.

MOYE, D. G., 1959, "Rock Mechanics in the Investigation and Construction of Tumut I Underground Power Station," *Engineering Geology Case Histories No. 3,* Symposium on Rock Mechanics, Geological Society of America.

MÜLLER, L., 1963, *Felsbau,* Vol. I. Stuttgart: F. Enke.

MÜLLER, L. (editor), 1974, *Rock Mechanics.* Wien-New York, N.Y.: Springer-Verlag.

National Academy of Sciences, 1974, *Advances in Rock Mechanics,* Vols. I and 2, Reports on Current Research, Proceedings Third Congress of the International Society for Rock Mechanics held at Denver, Colo.

OBERT, L., and W. I. DUVALL, 1967, *Rock Mechanics and the Design of Structures in Rock.* New York, N.Y.: John Wiley and Sons, Inc.

POUGH, F. H., 1976, *A Field Guide to Rocks and Minerals* (4th ed.). Boston: Houghton Mifflin Company.

PRAGER, W., and P. HODGE, 1951, *Theory of Perfectly Plastic Solids.* New York, N.Y.: John Wiley and Sons, Inc.

SCHULTZ, J. R., and A. B. CLEAVES, 1955, *Geology in Engineering.* New York, N.Y.: John Wiley and Sons, Inc.

SOKOLNIKOFF, I. S., 1956, *Mathematical Theory of Elasticity.* New York, N.Y.: McGraw-Hill Book Company, Inc. (476 pages).

SOUTHWELL, R. V., 1969, *An Introduction to the Theory of Elasticity and Plasticity for Engineers and Physicists.* New York, N. Y.: Dover Publications, Inc.

STAGG, K. G., and O. C. ZIENKIEWICZ (editors), 1968, *Rock Mechanics in Engineering Practice.* New York, N. Y.: John Wiley and Sons, Inc.

SZÉCHY, K., 1966, *The Art of Tunnelling.* Budapest: Académiai Kiadó.

TALOBRE, J. A., 1957, *La Mécanique des Roches.* Paris: Dunod.

TIMOSHENKO, S., and J. N. GOODIER, 1951, *Theory of Elasticity.* New York, N. Y.: McGraw-Hill Book Company, Inc.

TREFETHEN, J.M., 1959, *Geology for Engineers* (2nd ed.). New York. N. Y.: D. Van Nostrand Company, Inc.

WAHLSTROM, E. E., 1974, *Dams, Dam Foundations and Reservoir Sites.* Amsterdam: Developments in Geotechnical Engineering 6, Elsevier Scientific Publishing Company.

PART 2

ROCK EXPLORATION

CHAPTER 4

METHODS OF ROCK EXPLORATION

4-1. Need for Rock Exploration

In rock engineering, rock exploration is as important as is soil exploration in soil and foundation engineering.

Civil engineering practice has demonstrated that careful design of an engineering structure is not yet all that is needed for its safety and stability. Before any design and construction are contemplated, a comprehensive and careful exploration should be made of the environment of the structure to be built. The said applies especially to the structure-supporting geological material, viz., rock upon which the structural foundation is to be laid. In this respect, the advisability of engaging the professional services of a competent engineering geologist should not be overlooked. However, the responsibility for the safety and success of the structure-rock system is the responsibility of the design engineer.

The position of the geologist is not to make restrictive requirements for the engineer, but to furnish full information upon the basis of which the engineer can make the best and most economical design.

Such an attitude is certainly commendable, and should make for the best of relationships in the field.

4-2. Exploration Methods

Rock exploration of a construction site and its vicinity is usually made along the following seven lines, namely:

1. Geological exploration,
2. Hydrological exploration,
3. Geophysical exploration,
4. Thermal exploration,

5. Evaluation of rock as to its diminution and workability,
6. A study of procurement of rock for construction, and
7. In-situ and laboratory testing of strength properties of rocks.

4-3. Geological Exploration

Geological exploration is concerned with geological conditions of rocks of the terrain and the construction site, as well as with the various properties of rock. Of interest are geological structures; stratification; tectonic conditions such as faults and systems and nature of rock weaknesses and discontinuities; petrographic nature of rocks and their effect on supporting loads; and rock performance under the action of load, water, and temperature. The weathering process of rock, too, should not be overlooked. Blasting of rock may open fissures and cracks and facilitate influx of water into underground openings.

Design and construction in rock in earthquake-prone regions should conform with local aseismic design building codes.

Geological exploration also includes mapping of rock exposures, and the study of areal rock patterns on the ground as well as from aerial photographs.

To reveal rock structure and material at the site under investigation, core drillings or borings are made and rock cores are extracted. Study of rock cores enables the construction of graphic profile. The depth at which a recognizable rock stratum is encountered by a series of bore holes gives a clue to rock structure. Core drilling, however, is expensive. A less expensive method of logging is electrical logging. The latter can be utilized by using boreholes drilled by less expensive methods. Groundwater that saturates sedimentary rocks contains dissolved salts. A salt solution is a good electrical conductor. The resistivity of the groundwater at any desired depth is measured by special sensors. Such electrical measurements give an electric log that is widely used in correlation.

4-4. Hydrological Exploration

Hydrological exploration should furnish information on precipitation; surface runoff; floods and tide conditions; groundwater-carrying layers (their numbers, depth); position of the groundwater table and its fluctuation; groundwater flow (direction and velocity); location of springs, possibility of influx of water in rock excavations and underground openings. Also, groundwater must be tested for its aggressiveness with regard to soil, rock, and foundation construction materials.

Saturated rock, on the average, has less strength than a dry one.

One must always count upon an influx of water into an underground opening. Influx of water in tunnels has a detrimental effect on the safety of tunnel traffic, especially if cold air at a temperature below 0 °C has access into the tunnel. Water on the roadway pavement may freeze to form ice.

Because water seeks the path of least resistance to flow, there should be a maxim observed to provide, in every dewatering system of rock, for the most convenient way possible for water to drain and flow away freely toward an outlet. Or else, because one must always count upon the influx of water, the use of efficient pumping systems must be considered. Also, water flowing out from rock slopes influence rock slope stability. When discharging on a highway pavement during freezing periods, the ice formed may create a traffic hazard.

Movement of underground water brings about the dissolution of certain minerals in rocks (karst regions of limestones).

Migrating groundwater may wash out soft clays and other decomposed materials from joints and faults. Under varying humidity and alternate drying and wetting conditions, some rocks separate on their bedding planes or, in time, disintegrate completely.

In foundation and rock engineering, the chemical content of water is evaluated mainly from the point of view of its aggressiveness relative to structural foundation materials, rocks, and tunnel-lining materials.

4-5. Geophysical Exploration

Geophysical exploration is practiced to detect changes in the rock of some physical property—for example, specific gravity, magnetism, or the transmission or reflection of seismic waves.

Magnetometric surveys are most successful in regions where igneous rocks (containing magnetite) project into overlying sedimentary rocks by intrusion, or faulting, or as erosional hills beneath surfaces of unconformity.

Rock exploration in situ can be made quickly and relatively inexpensively by means of seismic methods. In seismic exploration, an artificial earthquake is brought about by exploding a buried charge. From the travel time of seismic waves, which are picked up at the ground surface by means of seismographs, the thickness and condition of various rock strata can be determined. Sometimes seismic methods are used to determine rock moduli. However, from the rheological point of view, a seismic test is a short-time test during which a rock performs rather elastically if no time is allowed to pass. In sustained long-term performance, rock exhibits a creep phenomenon.

4-6. Thermal Exploration

Thermal and hydrothermal exploration should furnish information about underground temperatures and thermal gradients in rock. This information is needed for design of ventilation and/or air conditioning systems for underground structures during their construction and service.

Also, this phase of exploration should pay attention to anticipated influx of gases into underground opening and subsurface structures. Gas exploration results may have a bearing on the capacity and efficiency of the ventilation and refrigeration systems and other kinds of machinery involving heat transfer.

Temperature variations may bring about spalling of a rock. Large variations in temperature induce thermal stresses in rock, especially in granite and other coarse-grained rocks, resulting in "peeling" of the rock surface. Such rocks are usually rejected in summer-winter climates for their use as a riprap material in protecting slopes against erosive wave action.

4-7. Evaluation of Rock as to its Diminution and Workability

Evaluation of rock as to its diminuity and workability is of great importance, and necessary for procurement and effective use of equipment for work in rock. This evaluation also has a bearing on overbreak, in making decisions as to whether or not rock bolting is needed, or if ring supports are to be used for supporting tunnel walls, and the like.

Rock crushing and grinding constitute an integral part of temporary industry.

Rocks differ greatly in their value for building purposes.

They vary markedly in their weathering qualities-resistance to atmospheric agents such as rain, frost, and wind.

Rocks vary in hardness; this factor affects the rate of drilling them and thus the cost.

The discussion about procurement of rock for construction is beyond the scope of this book. For in-situ and laboratory testing of strength properties of rocks refer to Sections 4-26 through 4-38 of this book.

4-8. Properties

Rock is a natural substance. As such, it has structural features which are not encountered in most other engineering materials.

The performance of soil and rock under the action of load, water, temperature, and tectonics of the earth's crust depends upon *physical and mechanical* (strength) *properties* of those materials.

One should distinguish between macro- and microscale properties of rock. Essentially, macroscale rock mass properties are properties of the compound, whereas microscale (small-scale) rock properties are those of the rock substance.

For safe and economical design of a structure on and in rock, adequate knowledge about the technological properties of rock is indispensible. Some of the main technological properties of rocks necessary to consider in the design of concrete dams, rock slopes, and underground structures are:

1. Unit weight,
2. Movements and deformations under load (absolute or relative, or strains),
3. Static and dynamic strength (compressive, tensile, bending, shear, punch, torsion) of dry and saturated rock
4. Angle of internal friction φ
5. Cohesion c
6. Young's modulus of elasticity E
7. Poisson's number $m = 1/\mu$ (μ = Poisson's ratio)
8. Natural or primary rock stress
9. Impact resistance
10. Resistance to mechanical abrasion (wear)
11. Creep
12. Dilatancy
13. Density against water absorption
14. Permeability to water
15. Resistance to weathering
16. Resistance to frost action in soil (freezing and thawing)
17. Resistance to chemical influences
18. Thermal properties (conductivity, heat capacity)
19. Electrical properties, and
20. Other possible properties.

Also, it can be brought out that some of the properties of rock materials may be elastic, plastic, viscous, or a combination thereof.

4-9. Physical Properties

The physical properties of rocks affecting design and construction in rock are:

1. Mineralogical composition, structure, and texture;
2. Specific gravity G
3. Unit weight γ
4. Porosity n
5. Void ratio e
6. Natural moisture content w
7. Saturation moisture content
8. Degree of saturation S
9. Permeability to water k
10. Chemical effects
11. Thermal properties
12. Electrical properties.

Mineralogical composition is the intrinsic property controlling the strength of the rock.

Although there exist more than about 2,000 kinds of known minerals, only about nine of them partake decisively in forming the composition of rocks. They are: quartz, feldspars, micas, hornblende, augite, olivine, calcite, kaolinite, and dolomite. Rocks containing quartz as the binder are the strongest, followed by calcite and ferrous minerals as the cementing agent "glue". Rocks with clayey binder are the weakest.

Rock Structure

According to KRYNINE and JUDD (60), the term "rock structure" applies to well-pronounced macroscopic features of the rock (columnar structure, for example). This term may signify special features of the rock, such as the position, arrangement, and attitude of a system of joints (open or closed); fractures; successive strata of rock formations; holes and/or cavities (small or large), gouge fillings, and the geologic features which determine that attitude such as stratification, faults, folds (anticline and syncline), and igneous intrusions, for example.

Rock Texture

The term "rock texture" refers to the arrangement of its grains or particles on a freshly exposed rock surface, easily seen by the naked eye. Texture is the appearance, megascopic or microscopic, seen on a smooth surface of a mineral aggregate, showing the geometrical aspects of the rock including shape, size, and arrangement.

One distinguishes between coarse-textured (coarse-grained) and fine-textured rock. A coarse-textured rock is one in which the large crystals or other kinds of grains can be seen easily by the naked eye.

A fine-textured rock is one whose grains cannot be seen without magnification.

An igneous rock whose coarse crystals are embedded into a fine-textured matrix is called porphyry.

Obviously, rock structure and texture affect the strength properties of the rock.

From a morphological point of view, rock texture may be grouped into three main groups, namely:

1. Homogeneous,
2. Nonhomogeneous or heterogeneous, and
3. Layered.

Specific gravity G of a material is defined as the ratio of the weight of a material to the weight of an equal volume of water. In other words, G is a number expressing how many times a material is heavier (lighter) than an equal amount of volume of water.

True or absolute specific gravity G of a soil and/or rock is the specific gravity of the actual soil and/or rock grains or solids:

$$G = \frac{W_d}{W_{\text{sat in air}} - W_w - W_{\text{sat in water}}} \tag{4-1}$$

$$G = \frac{\text{weight of rock (soil) particles in g}}{\text{(volume of soil particles) (1.00)}} = \frac{W_s}{V_s \cdot \gamma_w} = \frac{\gamma_s}{\gamma_w} \tag{4-2}$$

A rock mass may comprise several different minerals of varying absolute specific gravities, but its true or absolute specific gravity is the average determined from a representative rock specimen.

Specific gravity G is used in computing other soil and rock properties; for example, the porosity and void ratio of a rock and soil. Table 4-1 shows specific gravities of some minerals.

TABLE 4-1

Specific Gravity of Some Minerals

No.	Mineral	Specific gravity G
1	2	3
1	Biotite (black mica)	2.70–3.20
2	Calcite	2.71–3.72
3	Dolomite	2.80–3.00
4	Feldspar	2.50–2.80
5	Gypsum	2.20–2.40
6	Hornblende	2.90–3.50
7	Kaolinite	2.50–2.65
8	Montmorillonite	2.00–2.40
9	Muskovite (white mica)	2.76–3.00
10	Quartz	2.65
11	Talc	2.60–2.70

Bulk or apparent specific gravity G_b is the ratio of the dry weight W_d of rock to the weight W_w of volume of water equal to the *total* volume of rock including voids.

For water, its specific gravity is assigned the numerical value of $G = 1.00$.

$$G_b = \frac{W_d}{W_w} \tag{4-3}$$

where $W_w = W_{\text{sat in air}} - W_{\text{sat in water}}$

Unit Weight

The unit weight γ of soil and rock above the groundwater table is expressed as the ratio of the total weight of the rock (soil) in air W to the total volume V of the rock, all voids included:

$$\gamma = \frac{W}{V} \tag{4-4}$$

Rocks containing heavy minerals have higher unit weight than rocks with lighter minerals. Usually, igneous and metamorphic rocks have greater unit weight than sedimentary rocks. The more porous the rock is, the less is its unit weight.

The dry unit weight of rock (soil) γ_d is computed as:

$$\gamma_d = \frac{W_d}{V} = \frac{G}{1+e} \cdot \gamma_w = \frac{\gamma}{1+w} \tag{4-5}$$

where W_d = dry weight of rock specimen;
V = total volume of the specimen;
G = absolute or true specific gravity of soil, or average specific gravity of rock;
e = void ratio;
γ = moist unit weight;

$$\frac{1}{1+e} = \frac{V_s}{V} \tag{4-6}$$

V_s = volume of solids;
γ_w = unit weight of water = 1 g_f/cm^3 = 1 t_f/m^3;
w = moisture content, by dry weight, in decimal fractions.

The unit weight γ of a moist soil calculates as:

$$\gamma = (1-n) \cdot G\gamma_w + nS\gamma_w \tag{4-7}$$

where

$$S = \frac{w}{w_{sat}} = \frac{n_w}{n} = \frac{V_w}{V_v} = \frac{wG}{e} \tag{4-8}$$

S = degree of saturation, in decimal fractions
w = moisture content in rock;
w_{sat} = saturation moisture content;
n_w = relative pore volume at moisture content w;
n = porosity of rock in decimal fractions;
V_w = volume of water in rock;
V_v = volume of all voids (pores) in rock;
G = average specific gravity of rock material, and
e = void ratio.

The unit weight of solids by absolute volume is

$$\gamma_s = G \cdot \gamma_w \tag{4-9}$$

Saturated unit weight γ_{sat} of soil or rock is:

$$\gamma_{sat} = (1-n)G\gamma_w + n\gamma_w \tag{4-10}$$

where

n = porosity of rock or soil,
G = specific gravity,
γ_w = unit weight of water
$= 1\ g_f/cm^3 = 1\ t_f/m^3 = 9.81\ kN/m^3 = 62.4\ lb/ft^3$

When inundated, the rock and soil particles are subject to buoyancy. The buoyant or submerged unit weight γ_{sub} is expressed as:

$$\gamma_{sub} = (1-n)(G-1)\gamma_w = \gamma_{sat} - \gamma_w \tag{4-11}$$

Dry unit weights of some American rocks are compiled after GRIFFITH (35) in Table 4-2:

$$1\ lb/ft^3 = 0.157092\ kN/m^3 = 0.0160184\ t_f/m^3 = 16.01846\ kg_f/m^3$$

TABLE 4-2

Dry Unit Weight and Porosity Values of Some American Rocks
(after J. H. GRIFFITH, Ref. 35)

Type of rock	Bulk or apparent specific gravity	Dry unit weight t_f/m^3	Porosity n %
1	2	3	4
Igneous:			
Basalt	2.21 – 2.77	2.21 – 2.77	0.22 – 22.06
Diabase	2.82 – 2.95	2.82 – 2.95	0.17 – 1.00
Gabbro	2.72 – 3.00	2.72 – 3.00	0.00 – 3.57
Granite	2.53 – 2.62	2.53 – 2.62	1.02 – 2.87
Sedimentary:			
Dolomite	2.67 – 2.72	2.67 – 2.72	0.27 – 4.10
Limestone	2.67 – 2.72	2.67 – 2.72	0.27 – 4.10
Sandstone	1.91 – 2.58	1.91 – 2.58	1.62 – 26.40
Shale	2.00 – 2.40	2.00 – 2.40	20.00 – 50.00
Metamorphic:			
Gneiss	2.61 – 3.12	2.61 – 3.12	0.32 – 1.16
Marble	2.51 – 2.86	2.51 – 2.86	0.65 – 0.81
Quartzite	2.61 – 2.67	2.61 – 2.67	0.40 – 0.65
Schist	2.60 – 2.85	2.60 – 2.85	10.00 – 30.00
Slate	2.71 – 2.78	2.71 – 2.78	1.84 – 3.61

Porosity of Rock

Voids, i.e. pores and fractures in rock, are very important forms of non-uniformity of structure and texture of rocks.

Pores in rock are little, interconnected voids having connection to the air also.

The quality of rock porosity is characterized by the term "porosity". The degree of porosity depends upon the kind and structure of the rock.

Pores have a considerable effect on heat, gas, and water conduction and on the behavior of rock relative to its moisture content.

The presence of voids or pore spaces in rocks affects negatively its mechanical—viz., strength—properties. A small amount of porosity in the form of cracks has a large effect upon the deformation of rock. All polycrystalline materials, among them also rocks, are relatively porous depending upon their type, mineral composition, and mode of formation. Porosity is the result of internal stresses developed by only modest changes in stress and temperature; or as a result of imperfect sintering, exsolution of gas; or, as in compact crystalline rock, the opening of small fissures by internal stresses (20). Upon quick cooling of the magma or lava, whichever the case is, fine crystals form in the igneous rock, resulting in a relatively non-porous rock; whereas upon slow cooling, large crystals are formed. When gases are released upon cooling, very porous rock results.

The porosity of a sedimentary rock depends upon the cementing material ("glue") of the aggregate present, and the granulometric contents of the rock. Usually, the largest porosities are observed in sedimentary rocks.

The amount of voids in a rock and soil mass can be expressed in terms of porosity n, and as void ratio e.

Porosity n is the amount of voids based on total volume V of rock:

$$n = V_v/V = \frac{e}{1+e} = \frac{V-(W_s/G\gamma_w)}{V} \tag{4-12}$$

where V_v = volume of voids,

$e = \dfrac{n}{1-n}$ = void ratio,

W_s = dry weight of rock solids,

G = specific gravity of soil or rock,

γ_w = unit weight of water, and

$$V = V_s + V_v = V_s + V_a + V_w \tag{4-13}$$

= total volume

V_s = volume of solids in the rock specimen,
V_a = volume of air or gas in the rock specimen,
V_w = volume of water in the rock specimen.

Porosity values of some American igneous, sedimentary, and metamorphic rocks are compiled in Table 4-2.

In general, a cubic meter of *solid rock* excavation will make more than a cubic meter of fill or embankment, because there will remain void spaces between the rock fragments as placed in the fill. A *"swell"* of approximately 25% can be expected, depending upon the size and gradation of the rock fragments and the applied compactive effort.

Porosity values are most important in the study of the physical and mechanical properties of rocks. Porosity also affects the electrical properties of rocks, and porosity values aid in interpreting electric logs.

A small amount of voids, viz., small porosity in the form of cracks has a great effect on the deformation of rock. Closing and opening of cracks under the influence of stress makes the rock stress-strain relationship nonlinear. Sliding between crack surfaces brings about hysteresis in the stress-strain graphs.

Void ratio e is the ratio of the volume of voids V_v to the volume of solids V_s of the rock, expressed in decimal fractions. It is a dimensionless number which simply shows how many times there are more voids than solids in the rock:

$$e = \frac{V_v}{V_s} = \frac{n}{1-n} \tag{4-14}$$

The volume of solids V_s calculates as:

$$V_s = \frac{W_s}{G \cdot \gamma_w} \tag{4-15}$$

Water in Rock

In rock engineering, one must always count upon the influx of water into underground openings. Water always seeks the path of least resistance to flow.

Fractures and other rock defects are the easiest passages of ingress for groundwater in the rock. Hence, the extent and the geometry of the fracture and jointing system in rock of various lithology, topography, orientation, relation to surface streams and drainage pattern, and depth below ground surface should be helpful in interpreting the hydraulic regimen of, and permeability conditions in the rock.

Rock pores such as cavities, fissures, and fractures may contain a certain amount of water that may be free or bound (water film, for example).

As to the water film, it is present in the earth's crust in great quantity. This interstitial water cannot be drained mechanically or pumped away. The water film moves thermo-osmotically under the action of a thermal gradient (or sometimes under an electrical gradient) in rock through the smallest, thinnest voids such as fissures and seams. In due time, this category of water brings about alteration, weathering, and, thus, fragmentation of the rock. Hence, these alterations affect the stress and hydraulic field within the rock mass. Therefore, the study of moisture migration in fissured and fine porous rocks requires the consideration of the thermo-osmotic phenomenon, which phenomenon, in its turn, requires knowledge of the electrical parameters of the rock — dielectric constant, electrical conductance and resistivity as a function of structure, texture, porosity, and moisture content of the rock materials (53, 56, 57).

Moisture content w of rock is the weight of water W_w present in a rock expressed in percentage by oven-dry weight of the rock:

$$w = \frac{W_w}{W_s} \cdot 100 = \frac{W - W_s}{W_s} \cdot 100 \tag{4-16}$$

The dry weight of the solids of rock calculates as:

$$W_s = \frac{W}{1 + (w/100)} \tag{4-17}$$

where W = total weight of rock specimen, including moisture.

Most rock contains moisture of less than 1% to more than 35% in porous rocks (such as sandstone, for example). In many mines and other underground works, the rock is nearly saturated. Results are leaky tunnels and water seeping and dripping into the opening and out from fractures, faults, and joints (Fig. 5-5).

Alternate wetting and drying causes some rocks to expand and contract, thus affecting their properties. Also, riprap stones may become destroyed by alternate wetting and drying.

Moisture adversely affects electrical and electronic equipment, installations, and optical devices.

TABLE 4-3

Some Soil Physical Constants

Properties to be Determined by Test	Quantities to be Calculated for a Soil in its Undisturbed State	
	Description	Equation
1	2	3
1. γ = unit weight of soil	1. Unit weight of soil skeleton or dry unit weight	(a) $\gamma_d = \frac{\gamma}{1+w}$ (b) $\gamma_d = \frac{G}{1+e}\gamma_w$
2. w = moisture content by dry weight	2. Porosity	$n = 1 - \frac{\gamma_d}{G\gamma_w}$
	3. Void ratio	$e = \frac{n}{1-n} = \frac{G\gamma_w - \gamma_d}{\gamma_d}$
3. G = specific gravity of soil particles	4. Relative volume of voids	$n = \frac{e}{1+e}$
	5. Relative volume of solids	$n_s = \frac{1}{1+e}$
4. γ_w = unit weight of water	6. Relative volume of water in soil	$n_m = nS$
	7. Soil moisture content by dry weight	$W_w = w\gamma_d$
	8. Soil moisture content by volume	$V_w = \frac{W_w}{\gamma_w}$
	9. Moisture content upon full saturation	$w_{sat} = \left(\frac{1}{\gamma_d} - \frac{1}{G\gamma_w}\right)\gamma_w$
	10. Degree of saturation	$S = \frac{w}{w_{sat}} = \frac{w\gamma}{n(1+w)\gamma_w}$
	11. Void ratio upon full saturation	$e = wG$
	12. Volume of air in a unit volume of soil	$n_a = (1 - S)\,n$
	13. Saturated unit weight of soil	$\gamma_{sat} = \frac{G+e}{1+e}\gamma_w$
	14. Submerged (buoyant) unit weight	$\gamma_{sub} = \frac{G-1}{1+e}\gamma_w$

Degree of saturation S is determined as:

$$S = \frac{V_w}{V_v} = \frac{w}{w_{sat}} = \frac{n_w}{n} = \frac{w \cdot \gamma}{n(1+w)\gamma_w} = \frac{wG}{e} = \frac{wG(1-n)}{n} \tag{4-18}$$

The functional relationships between various physical properties of soil or rock are summarized in Table 4-3. The volumetric and gravimetric relationships of soil or rock are illustrated graphically by means of a phase diagram as shown in Fig. 4-1. The phase diagram illustrates absolute and relative volumetric and gravimetric proportions of solids, water, and air in a unit of volume of soil or rock.

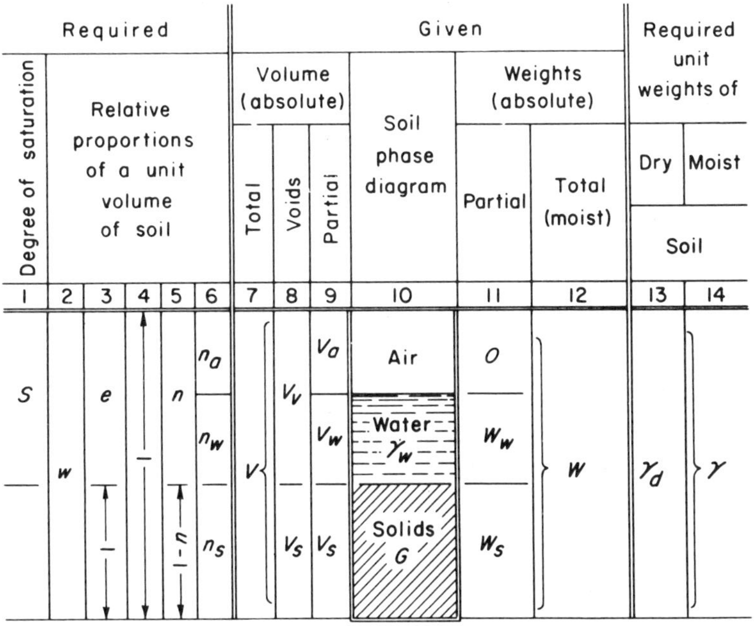

Fig. 4-1
Phase diagram illustrating absolute and relative volumetric and gravimetric proportions of solids, water, and air in a unit mass of soil

Permeability

Rock porosity and permeability to water are important physical properties in rock engineering.

Permeability is defined as the property of a porous material that permits the passage or seepage of fluids, such as water and/or gas, through its interconnecting voids.

The resistance to flow depends upon the type of the rock (soil), the geometry of the voids of rock (size and shape of the voids), and the surface tension of water (temperature and viscosity effects).

Although theoretically all rocks (soils) are more or less porous, in practice the term "permeable" is applied to rocks (soils) that are porous enough to permit flow of water through it. Conversely, rocks that are permeated with great difficulty are termed "impermeable".

1. Gravitational or free water drains from the voids of the rock by gravity.
2. Water film does not drain freely.
3. Water also may resist drainage by gravity if it is held in the soil by surface tension forces (capillarity).
4. Drainage lowers the position of the groundwater table.

Water seeps through soil at velocities usually well below the critical value at which turbulence appears (Fig. 4-2). The total discharge Q through a gross cross-sectional area A during time t is given by DARCY's law $v = ki$, namely:

$$Q = vAt = kiAt \quad [\mathrm{m}^3] \tag{4-19}$$

where $k = v/i = \tan \alpha$ = coefficient of permeability of the rock (soil) (it expresses the degree of permeability of rock or soil to water); and

$i = h/L$ = slope of the pressure gradient, or the pressure head h lost in a unit of distance.

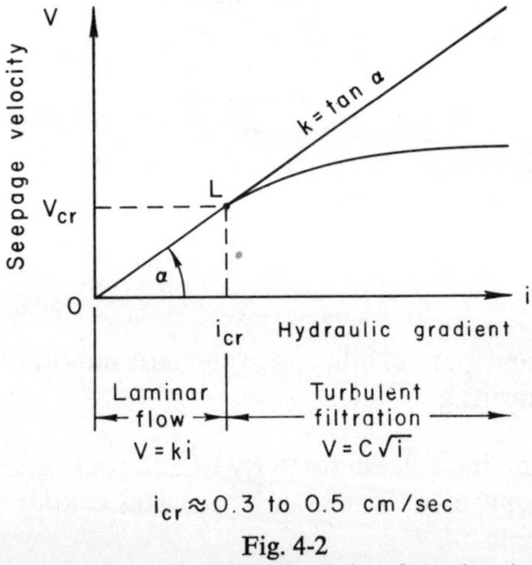

Fig. 4-2
Laminar region of DARCY's law from 0 to i_{cr}

The coefficient of permeability is a function of rock type, pore size, entrapped air in the pores, temperature in rock, and viscosity of water.

The order of magnitude of the coefficient of permeability of various types of soil to water is shown in Fig. 4-3.

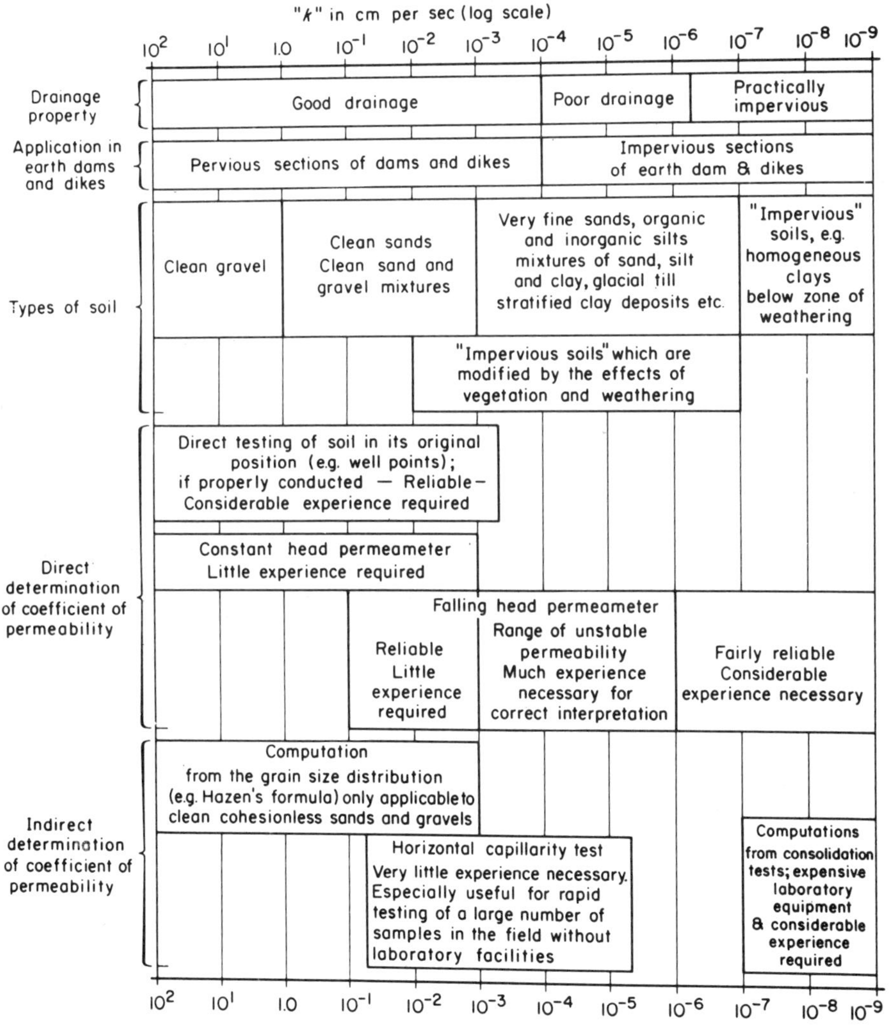

Fig. 4-3
Coefficient of permeability
(After CASAGRANDE and FADUM, Ref. 23)

Leakage in rock endangers the safety of underground openings.

In practice, the solution of drainage is frequently complicated because of the difficulty of providing satisfactory outlets in rock for the collection of water.

Relative to groundwater flow through rocks, the flow takes place through the fissures, jointing, and other defects of rock. Here, one must deal with fissure water in basalt, diabase, granite, gneiss, quartzite, limestone, sandstone and shale, for example.

Generally, the groundwater movement through fissures takes place according to the same laws as through the voids of soil, but with some difference. For example, in a porous medium like soil, water spreads out and distributes through the interconnecting pores in all directions, whereas in a fissured rock, water spreads and moves only along the directions of the fissures and joints.

While pore size in soil varies within relatively narrow limits, the size of fissures and joints may vary within very wide limits — from hair cracks to several meters in width.

Because of this irregularity, permeability of fissured rocks is said to be nonuniform. Also, expansion and contraction of fissures cause nonuniform and turbulent flow. Fissures also change size by erosion processes, and during the flow processes calcites and silicates may be washed out at one place and deposited at another place.

The correct evaluation of distribution of fissures and fissure water is possible only if based on:

1. Careful observation of size of fissures and their frequency distribution,

2. Research on transmissibility of water through rock; and

3. Rational systematization and classification of rock fissures and underground fissure water.

It should be distinguished between

1. *Tectonic fissures*—formed upon the formation of the structure of the rock;

2. *Lithogenetic fissures*—formed upon the formation of rock; and

3. *Weathering fissures*—brought about by weathering processes of rocks by water, wind, glaciers, and temperature.

Influx of water in underground openings and foundation pits requires its removal by pumping and/or drainage facilities.

The influx of water into the Bologna-Florence tunnel was up to 1200 liters per second, and into the St. Gotthard tunnel up to about 230 liters per second.

Influx of water in foundation pits and underground openings in rock as well as in soil frequently presents problems that are difficult to cope with. Also, if the rock near the toe of a slope is already stressed to near its point of failure, the added fissure and joint-water pressure may be large enough to bring about the failure of the slope.

The water seeping into the rock also pulls air with it into the depth, pushing air bubbles and air cushions ahead through the voids, thus promoting the air exchange and circulation within the rock.

Sometimes underground works affect the regimen of the surface courses: springs may become dry, and streams and lakes may disappear.

Stratigraphy is the basis for exploring hydrological conditions in rock at a site. Hydrological studies involve stream drainage pattern locations of springs. Also, sound knowledge about the physical properties of the rocks, but especially sedimentary rocks, are essential.

Chemical Action

Fissures, joints, and fault zones can be invaded by surface waters or by hot solutions from below, or by both. Hence, chemical alterations in such zones is likely to take place.

The chemical action of rainwater which enters the rock from the surface is referred to as the *chemical weathering*. Hot water (magmatic water) entering the voids in rock from great depth is said to bring about *hydrothermal alteration*. Hydrothermal agents and carbon dioxide are the main factors altering the feldspars of igneous and metamorphic rocks into kaolin and other clays.

Hardness of water — viz., the contents of various dissolved mineral salts, especially those of calcium and magnesium in groundwater—is of some interest to rock engineering. On the average, water is considered to be hard whose degree of hardness is pH < 7, containing free carbon dioxide, calcium chloride, chlorate, magnesium salts, and sulfates.

A soft water is one containing little dissolved electrolytes. It occurs in igneous rocks and glacial formations. Generally, the softness of water increases with the increase in coldness of the climate.

In foundation and rock engineering, the chemical effects of water are evaluated mainly from the point of view of the aggressiveness of water with respect to foundation materials, rocks in underground openings, tunnel-lining materials, and tunnel rock-bolt anchorages.

Depending upon the mineralogical composition of rocks, movement of underground water in rocks may result in dissolution of certain soluble mineral components and in the formation of karsts. For example, rocks containing pyrite FeS_2 and marcasite FeS_2 (white iron pyrite), may form sulfuric acid which is a strong solvent of calcite and magnesium salts.

In limestone, openings of solution channels and cavities are brought about by the decomposition of calcium carbonate produced by water containing dissolved carbon dioxide.

Thus, chemical actions may bring about changes in the mechanical properties of the rock.

In some instances, the groundwater is relatively acid. Humus and sulfur-containing water are detrimental to concrete even at relatively low concentrations.

Thus, the chemical alterations of rocks brought about by the solution depends upon the chemical composition of the solution and the type of rock, as well as upon the depth and temperature at which the changes take place. Therefore, because of chemical decay, presence of air, and the dissolution of certain rock-forming minerals by water and other agents in rock, the chemical nature of the groundwater must be investigated carefully.

Carbonic acid is a relatively weak acid, but it attacks concrete, and dissolves limestone and dolomite by removing calcium and magnesium from these rocks.

Sodium chloride (NaCl) is found in seawater and in many rocks. Sodium chloride solutions are not too dangerous as compared with solutions of magnesium sulfate ($MgSO_4$).

The salt termed sodium sulfate (Na_2SO_4) occurs in solution in seawater and in some mineral waters.

Calcium sulfate gypsum ($CaSO_4$) is very aggressive, and magnesium sulfate ($MgSO_4$) is even more detrimental than gypsum. Hence, rock formations containing gypsum, rock salt, and several other soluble minerals are subject to a rapid process of alteration, especially when changes in groundwater level fluctuation take place.

The factor of the possibility of swelling pressures of clay minerals to occur within the joints of a rock mass because of change in their moisture content is a very important factor in the stability of underground openings and rock slopes.

Gases

Rocks and building stones exposed to atmosphere can be detrimentally affected by atmospheric gases. Also, in-situ rocks can be seriously affected by underground gases.

The occurence of gases in underground openings depends upon the geological formations of the rock. Harmful gases may be encountered mainly in weathered, fissured, and jointed igneous rocks.

One of the most harmful gases is carbon dioxide (CO_2). It usually occurs in igneous rocks in the proximity of coal formations and other deposits of organic nature. In fault zones, the gas flows upward from the magma through faults, fissures, and other suitable passages.

The presence of carbon dioxide in underground works requires great safety precaution.

Usually carbon monoxide (CO) occurs in the proximity of coal deposits, as a result of methane and/or coal dust explosions.

Methane or marsh gas (CH_4) is encountered in the vicinity of coal and oil fields. It is also brought about by the decay of organic matter. Prevented from escape into the atmosphere, the gas can migrate long distances through the fissures of the rock.

Methane influx from a shale formation into the Great Apennine tunnel (Bologna-Florence) resulted in a great fire (86).

Sulfur dioxide (SO_2) and sulfur trioxide (SO_3) are harmful gases, also. When combined with atmospheric moisture, sulfuric trioxide forms sulfuric acid.

Sulfuric gases acting on calcite result in sulfate. As a result, rock scaling will take place. Therefore, under such conditions, limestone and marble do not resist weathering too well.

The inflammability of hydrogen (H_2) always involves the hazards of explosion and fire.

Nitrogen (N) is not inflammable. This gas is encountered in young eruptive rocks, or as the product of decay of organic matter of floral origin.

The aforegoing discussion about gases brings to the fore that it is important to explore gas exfiltrations, gas outbursts, and rock temperatures. Gas and temperature are significant factors for the safety and health of workmen and miners.

4-10. Electrical Properties

Most rocks are dielectrics, hence subject to dielectric constant measurement. This and other electrical properties depend upon the type of rock, its structure and texture, porosity, and moisture content (56).

Electrical properties of rocks are of interest for a number of practical purposes.

Electrical properties such as dielectric constant and electro-conductance are of great value in geo-electrical prospecting of groundwater resources in soil as well as in rock; in well-logging; in interpretation of electrical logs of soil and/or rock; in electrical sounding and profiling of rock; and in electrical prospecting for minerals.

The resistivity of rocks varies from 10^{-2} to 10^{16} ohm · cm.

4-11. Thermal Properties of Rocks

Heat and temperature are of special interest in rock engineering in connection with design and construction of tunnels and other kinds of underground openings (vertical shafts, power plants, shelters, storage spaces, and mines, for example). Also, it is known that an increase in temperature lowers rock strength and increases ductility. Therefore, knowledge about the thermal conditions in rock and of the geothermal gradient is of great technical and economical importance. This obtains especially with respect to observing health conditions and heat endurance limits for workmen; to designing efficient underground ventilation and air-conditioning systems; to choosing an appropriate method of construction; or to designing soil and rock artificial freezing and thawing facilities.

The thickness of the overburden cover of geological materials above an underground opening also influences the temperature in the opening.

Changes in temperature in rock may bring about rock exfoliation.

Heat

Heat is a transient form of energy, a thermal potential for translocation soil or rock moisture along a thermal gradient (53, 57).

Heat has a profound effect upon matter. Heat in the ground tends to expand and contract rock, water, gas, and air. Upon the addition of heat, solids and liquids increase in volume in most cases. Upon removal of heat, they contract. Heat has only one measurable property-temperature. Temperature is the measure of the intensity of heat energy. One of the characteristics of heat is its tendency to move.

When different parts of a rock are at different temperatures, heat flows from points at higher temperature to points at lower temperature. Because of nonuniform warming of the various geological strata of the earth's crust, heat exchange takes place constantly between these layers.

Change in temperature with depth within a stratum of rock and/or soil takes place in conformance with natural laws that are more complex than under usual, ordinary conditions.

The natural course of the air temperature follows the periodic oscillations of the atmospheric conditions, viz., climate.

The daily or diurnal temperature variations in soil can be measured to approximately 1 m below the ground surface. The monthly temperature variations in soil can be detected to a depth of about 7 to 10 m. The annual temperature variations extend down to about 20 m below the ground surface.

It is known that the temperature in rock formations of the earth's crust at shallow depths below the ground surface is a function of the temperature of the atmosphere above the ground surface. At a depth of about 25 to 30 m, the temperature is constant year round; it is about equal to the mean annual air temperature. However, deeper than about 30 m, the temperature starts to increase proportionally with depth.

The rate of increase in temperature with depth is expressed by a *geothermal step*. It is defined as the vertical distance along which there is a temperature increase of 1 °C. The inverse of the geothermal step is the *geothermal gradient*. It expresses the temperature increase for every 1-m depth.

The geothermal gradient is influenced by the width of water passages in the rock, distribution of the passages, temperature of the groundwater, extent and duration for filling the voids, velocity of flow of the fissure water, thermal conductivity of the rock, temperature of the trapped gas in rock, and other possible factors.

Tunnels driven through mountains made up of porous rocks cool considerably upon receiving meltwaters from snow and ice, or cold precipitation waters at temperatures somewhat above 0 °C.

In some mines, the rock temperature ranges from about −18 °C to +60 °C.

The temperature of the rock prevailing in the interior of the mountain depends upon the following factors (1):

1. Position of the geoisotherms under the mountain ranges (geothermal step),
2. Soil temperature of the ground surface above the tunnel,
3. Thermal properties of the rock,
4. Hydrological conditions above and below the ground surface, and
5. Position and elevation of the tunnel.

Table 4-4 shows overburden thickness and temperature in some tunnels (1,86).

TABLE 4-4

Overburden Thickness and Temperature in Some Alpine Tunnels

Tunnel	Length m	Over-burden thickness m	Max. temperature °C	Geo-thermal step m/°C	Rock
1	2	3	4	5	6
Albula	5,886	750	11–52	49	Granite
Apennine	18,500	2,000	64	—	—
Arlberg	10,250	715	18.5–34	38.6	Gneiss with mica granite, slate
Gotthard	14,998	1,752	40.4	47	
Karawanken	7,976	916	15.0	144	—
Lötschberg	14,605	1,673	34	45	Granite and slate
Mont Cenis	12,236	1,610	29.5	58,4	Sandstone and limestone
Simplon	19,729	2,135	55.4	37	—
Tauern	8,551	1,567	23.9	49	Granite, gneiss, and mica schist

Some of the more important thermal properties of rocks are:

1. heat capacity,
2. latent heat of fusion,
3. thermal conductivity,
4. thermal diffusivity or temperature conductivity, and
5. thermal expansion and contraction.

Heat Capacity

Among the important thermal properties of a material is its heat capacity, Mass heat capacity c_m of a substance is the actual amount of heat energy Q necessary to change the temperature of a unit mass, say 1 kg, ob the substance by one degree:

$$c_m = Q/\Delta T \quad [\text{Cal/(kg) (°C)}] \tag{4-20}$$

where c_m = unit of heat capacity for the mass m of a substance,
$\Delta T = T_1 - T_0$ = rise in temperature,
T_o = initial temperature, and
T_1 = final temperature ($T_1 > T_o$).
Cal = large or kilo-calorie = 1000 cal (small or gram-calories)
kg = kilogram

Numerically (but not dimensionally), the magnitude of the heat capacity is equal to the dimensionless value of specific heat.

Specific Heat

The specific heat of a substance is the dimensionless ratio of the heat capacity of a substance to the heat capacity of water.

In heat transfer problems in soil and rock engineering, the concept of volumetric heat capacity c_v of a substance is defined as the amount of heat necessary to change the temperature of a unit of the substance by one degree.

The relationship between the mass heat capacity c_m and volumetric heat capacity is:

$$c_v = c_m \cdot \gamma_d \quad [\text{Cal/(m}^3\text{) (°C)}] \tag{4-21}$$

where γ_d = dry unit weight of material, in kg_f/m^3.

One observes that the coefficient of the volumetric heat capacity of soil or rock depends upon its unit weight and its moisture content.

At freezing-point temperatures, the values of mass heat capacity are (57):

Water:	$c_{mw} = 1.0$	[Cal/(kg) (°C)]
Ice:	$c_{mi} = 0.5$	[Cal/(kg) (°C)]
Rock and dry soil mineral particles:	$c_{mr} = 0.2$	[Cal/(kg) (°C)]

The relationship between the volumetric heat capacity c_w and dry unit weight for various soil moisture contents w in percentages are:

For unfrozen soil or rock:

$$c_{vu} = \gamma_d \left[c_{mr} + \frac{(c_{mw})\,(w)}{100} \right] \quad [\text{Cal}/(\text{m}^3)\,(^\circ\text{C})] \tag{4-22}$$

For frozen soil or rock:

$$c_{vf} = \gamma_d \left[c_{mr} + \frac{(c_{mi})\,(w)}{100} \right] \quad [\text{Cal}/(\text{m}^3)\,(^\circ\text{C})] \tag{4-23}$$

In Table 4-5, mass heat capacity values of some minerals and rocks are compiled (24, 57, 83).

TABLE 4-5

Mass Heat Capacity of Minerals and Rocks

Substance	Temperature °C	Mass heat capacity c_m Cal/(kg) (° C)
1	2	3
Calcspar, $CaCO_3$	0– 50	0.188
Calcspar, $CaCO_3$	0–100	0.2005
Hornblende	20– 98	0.195
Mica (Mg)	20– 98	0.2061
Mica (K)	20– 98	0.2080
Basalt (fine, black)	12–100	0.1996
Dolomite	20– 98	0.222
Gneiss	17– 99	0.196
Granite	12–100	0.192
Limestone	15–100	0.216
Marble	0–100	0.210
Quartz sand	20– 98	0.191
Sandstone	–	0.22

Latent Heat of Fusion

When water changes its phase from liquid to solid, heat is liberated (latent heat of fusion). When ice melts, it absorbs heat from the air in contact with it. The amount of heat energy necessary to make this change isothermally, i.e. with no temperature change, is termed the latent heat of fusion, L, in Cal/kg. Because 1 kg of water liberates (or absorbs) $L = 80$ Cal/kg of latent heat, the total latent heat of fusion, Q_L, of $w\%$ water per 1 m³ of soil or rock material is:

$$Q_L = (\gamma_d)\,(w)\,(L) \quad \left[\frac{\text{Cal}}{\text{m}^3}\right] \tag{4-24}$$

where γ_d = dry unit weight of the material; and
w = moisture content of soil or rock by dry weight, in decimal fraction.

Notice that the amount of latent heat of fusion depends upon the dry unit weight of the rock (soil) material, and its moisture content w.

Thermal Conductivity

The total amount of heat Q transferred in the steady state by conduction through unit area perpendicular to flow ($A = 1.0$) during a time interval t under a temperature gradient of $(T_1 - T_2)/x$ is calculated as:

$$Q = K\,A\,t\,(T_1 - T_2)/x \quad [\text{Cal}] \tag{4-25}$$

where K = coefficient of thermal conductivity of rock, in Cal/(m · h · °C); here
Cal = kilo-calorie (large calorie)
$T_1 - T_2$ = temperature difference in °C between two points in the material x distance apart, and
$(T_1 - T_2)/x$ = temperature gradient in °C/cm.

Thermal conductivity K is not necessarily a constant but, in fact, is a function of temperature for all phases, and in liquids and gases depends also upon pressure. When the thermal conductivity is constant, a linear thermal gradient exists.

The coefficient of thermal conductivity is very sensitive to rock or soil type and its moisture content.

Thermal Diffusivity

The term thermal diffusivity is also known by the term temperature conductivity. When dealing with variable heat flow, in thermal calculations one uses the quantity thermal diffusivity α. It is connected with the thermal conductivity K as:

$$\alpha = \frac{K}{(c_{mr})(\gamma_d)} = \frac{K}{c_v} \quad \left[\frac{m^2}{h}\right] \tag{4-26}$$

where c_{mr} = mass heat capacity of rock,
c_v = volumetric heat capacity, and
γ_d = dry unit weight of rock (kg_f/m^3).

Thus, thermal diffusivity is a measure of the rate at which a change in temperature spreads through a body.

Equation (4-26) shows that at high moisture content, the thermal diffusivity α decreases, because the product $(c_{mr})(\gamma_d)$ in the denominator increases more rapidly than the thermal conductivity K.

Thermal conductivity and diffusivity values of some geological materials are compiled in Table 4-6 (24, 34, 57, 83).

TABLE 4-6

Thermal Conductivity K and Diffusivity α Values of Some Geological Materials

(Refs. 24, 34, 57, 83)

Material	K Cal/(m · h · °C)	α m²/h
1	2	3
Clays	0.21	0.0035
Sands (dry)	0.23	0.00072
Granite	2.08	0.0027
Limestones	1.58	0.0022
Sandstones	1.4	0.0022
Shales	0.8 – 1.25	0.0020

4-12. Coefficient of Thermal Expansion and Contraction

This topic should be of interest to geotechnical engineers because temperature has a marked effect on materials and engineering design. Temperature changes induce thermal stresses and strains in construction elements. Besides, one should be cognizant that upon freezing of soil and/or rock water, ice expands, while soil and rock mineral particles generally contract.

The ratio of the increase in length (linear coefficient), area (superficial), or volume (cubical) of a body for a given rise in temperature (usually from 0 °C to 1 °C) to the original length, area, or volume is known as the coefficient of thermal expansion. The three coefficients are approxiamately in the ratio of 1:2:3.

If the temperature of a body is warmed (or cooled) from a temperature T_1 to T_2 ($\Delta T = T_2 - T_1$), the change ΔL (expansion or contraction) of the initial length L of a specimen of the body calculates as:

$$\Delta L = \alpha_t \cdot L\,(T_2 - T_1) = \alpha_t \cdot L \cdot \Delta T \tag{4-27}$$

Its new length L_t is approximately equal to:

$$L_t = L + \Delta L = L(1 + \alpha_t \cdot \Delta T) \tag{4-28}$$

Its new volume V_t is:

$$V_t = V(1 + \beta \cdot \Delta T) \tag{4-29}$$

where V = initial volume,
α_t = coefficient of linear thermal expansion, expressed in units of cm per cm per degree of temperature change as

$$[(\text{cm}/\text{cm})/^\circ\text{C}] = [1/^\circ\text{C}]$$

β = coefficient of volumetric (cubical) thermal expansion $[1/^\circ\text{C}]$.

Here,

$$\alpha_t = \frac{1}{L} \cdot \frac{dL}{dT} \quad [1/^\circ\text{C}] \tag{4-30}$$

or

$$\alpha_t = \frac{\Delta L}{L} \cdot \frac{1}{\Delta T} \tag{4-31}$$

and

$$\beta = \frac{1}{V} \cdot \frac{dV}{dT} \quad [1/^\circ\text{C}] \tag{4-32}$$

or

$$\beta = \frac{\Delta V}{V} \cdot \frac{1}{\Delta T} \tag{4-33}$$

The coefficient α_t of linear thermal expansion (contraction) for solid bodies is the relative change in length dL/L per degree C increase (decrease) in temperature (ratio of dL per degree to the length L at 0 °C).

The coefficient β of volumetric thermal expansion (contraction) for solids is the relative change in volume dV/V per degree increase (decrease) in temperature (ratio of dV per degree to the volume of 0 °C).

Water has a volumetric (cubical) expansion coefficient of (40):

$$V_t = V[1 + a(\Delta T)^2 + b(\Delta T) + c(\Delta T)^3 + \ldots] \tag{4-34}$$

where $a = -6.43 \times 10^{-5}$,
$b = 8.50 \times 10^{-6}$, and
$c = 6.79 \times 10^{-8}$.

For homogenenous bodies such as most of the sound rocks that expand equally in all directions, the coefficient β of volumetric thermal expansion is three times the linear one:

$$\beta = 3\alpha_t \tag{4-35}$$

One notices that thermal expansion is proportional to the rise (fall) in temperature.

Some coefficients of linear thermal expansion α_t of some geological materials are compiled in Table 4-7 (34).

TABLE 4-7

Average Coefficients α of Thermal Expansion of Some American Rocks at Room Temperature to 100° C

(after GRIFFITH, Ref. 34)

Name of Rock	α 1/° C
Igneous	
Granite series	$34 \times 10^{-7} — 66 \times 10^{-7}$
Basalt series	$22 \times 10^{-7} — 35 \times 10^{-7}$
Diabase	$31 \times 10^{-7} — 35 \times 10^{-7}$
Gabbro	$20 \times 10^{-7} — 30 \times 10^{-7}$
Sedimentary	
Limestones and dolomites	$24 \times 10^{-7} — 68 \times 10^{-7}$
Sandstones	$36 \times 10^{-7} — 65 \times 10^{-7}$
Metamorphic	
Gneisses	$34 \times 10^{-7} — 44 \times 10^{-7}$
Marbles	$34 \times 10^{-7} — 51 \times 10^{-7}$
Quartzites	$60 \times 10^{-7} — 61 \times 10^{-7}$
Schists (crystalline)	$34 \times 10^{-7} — 43 \times 10^{-7}$
Slates	$45 \times 10^{-7} — 49 \times 10^{-7}$

It should be noticed that the coefficients (α_t) of layered soils and rocks are anisotropic relative to thermal expansion (contraction) because of the anisotropy of the layered soil (rock) systems. Also, with increase of all-sided confining pressure, the linear and volumetric thermal coefficients of *expansion* decrease.

The relationship between thermal expansion and various other factors such as composition of soils and rocks, density, moisture content, and others are practically unexplored as yet.

Mechanical Properties of Rocks

4-13. General Properties

Generally, the mechanical properties of a material characterize its reaction to the effect of the force field of its environment. Particularly, the mechanical properties of rocks depend upon:

1. The nature of the rock substance,
2. The stratigraphy of the rock in situ,
3. Rock defects, and
4. Testing methodology.

The most important mechanical properties of rocks that must be thoroughly investigated when designing foundations, hydraulic structures, and underground openings relative to failure of a geologic material are:

1. Hardness,
2. Durability,
3. Permeability to water,
4. Elasticity,
5. Plasticity,
6. Deformability, and
7. Strength.

The mechanical properties of a rock material are elasticity, plasticity, viscous deformations, and combinations thereof.

However, for safe, effective work in rock, the mechanical properties of most concern are the strength properties relative to the competency of rocks in underground opening and open-cut excavations. These rock properties vary with depth of the rock formation.

Hardness

Hardness (written H) of a mineral and rock is the resistance to abrasion. Every mineral and every rock have a hardness or range of hardness that ultimately depends on the strength of chemical bonds.

To rate hardness of minerals and rocks, the empirical MOHS' hardness scale is used (Table 4-8). The hardness may be determined by scratching one mineral with another. Table 4-8 shows an arbitrary numerical scale or rating of relative hardness based on common minerals. The higher rating numbers represent harder substances. Each material in the MOHS' scale of hardness scratches the previous one by the following one.

TABLE 4-8

Moh's Scale of Hardness of Minerals

Number of relative hardness scale or rating H	Mineral	Chemical composition	Remarks
1	2	3	4
1	Talc	$Mg_3Si_4O_{10}(OH)_2$	Softest; can be scratched by fingernail.
2	Gypsum	$CaSO_4 \cdot 2H_2O$	Can be scratched by fingernail.
3	Calcite	$CaCO_3$	A copper penny or a brass pin can scratch calcite.
4	Fluorite	CaF_2	Fluorspar. May be scratched easily by a steel point.
5	Apatite	$Ca_5F(PO_4)_3$	Any of the calcium phosphate minerals. Can be scratched by a knife. A window glass may be rated as $H = 5.5$ on the hardness scale.
6	Orthoclase (feldspars)	$KAlSi_3O_8$	Can be scratched by a knife blade of a good-quality steel; a hardened steel file may be rated as $H = 6.5$
7	Quartz	SiO_2	Scratches steel, glass, and all of the minerals whose $H < 7$.
8	Topaz	$Al_2SiO_4(F, OH)_2$	Great hardness. A valuable jewelry stone.
9	Corundum	Al_2O_3	Harder than any other natural mineral except diamond. An important industrial abrasive and refractory. Has many gem varieties, among them ruby and sapphire.
10	Diamond	C	The hardest substance known; not all diamonds are of the same hardness, however.

Sometimes hardness is used as a strength criterion in rocks.

The strength of sandstone rock increases with increasing amount of quartz.

The hardness of the minerals contained in a composite rock reflects to a certain extent the mechanical properties of a rock.

For example, based on his experimental results, Price (75) showed the effect of quartz content in sandstone with a calcite matrix, and in siltstone with a clay matrix, on the uniaxial compressive strength of these materials, namely: the strength of the rock materials tested increased considerably with increasing quartz content.

In general, excavation of solid, hard rock for its removal, which requires blasting, is more difficult and therefore more expensive than excavation of soil materials.

4-14. Durability

Durability of rock is a relative term. This property depends upon the nature of the rock environment—such as the climate and atmosphere, for example—and the amount of exposure of rock or building stone in the structure (viz., weathering).

Also, the rate of solution action on limestone, dolomite, and cement grout; resistance of a rock to frost action (freezing and thawing); rate of weathering; and porosity of rock may be correlated with durability.

According to the National Bureau of Standards [Eshbach, (27)], the estimates for the life of rock, viz., building stone, are:

Ohio sandstone, best variety	1 year to many centuries
Limestone	20 to 40 years
Marble, coarse	40 years
Marble, fine	50 to 200 years
Gneiss	50 years to many centuries
Granite	75 to 200 years

Generally, sound sandstone is durable, and is rated as having good fire resistance.

Therefore, by means of petrographic and other appropriate analyses, and by freeze-thaw tests, the resistance of rocks to weathering and chemical aggressiveness by groundwater and precipitation should be ascertained.

4-15. Elasticity

Elasticity is a universal property of an ideal material. Every solid material deforms under the action of a load, viz., stress. To every kind of stress there is a corresponding strain. If the stress is not too large, the strained ideal material will recover its unique, natural state, the original shape and size to which the material returns when the external loads are removed. All stresses, strains, and displacements are measured from this natural state; their values are taken as zero in this state. The property of recovering from strain of a material is called *elasticity*. If a material recovers completely, it is called perfectly elastic. If the material does not recover completely, the strain that remains when the stress is removed is called *permanent set,* and the material is said to be in an elastic state.

A relatively small but easily measurable part of the total deformation of solid bodies under load is said to be of an elastic nature.

Relative to small reversible distortions or deformations, ordinary solids have *isotropic elasticity* (quartz and diamond, for example).

Depending upon how closely a rock approximates the ideal material, the concept of elasticity applies also to rocks. Thus, one gets the notion that in practice, the property of elasticity of rocks depends upon their continuity, homogeneity, and isotropy.

Because of the manifold factors involved in rock strength, in rock engineering practice when designing in rock and performing stability analyses of rocks, certain idealizations and assumptions as to the nature of rock as a construction material are usually made. The following are some of the most important and most frequently made idealizations:

1. The rock is assumed to be a continuous, homogeneous, linear-elastic, isotropic material. The group of materials whose mechanical property does not depend upon the direction is said to be isotropic.

2. This material obeys HOOKE's law of proportionality between stress and strain; that is, the strains are linear functions of stresses.

3. The deformations (strains) of the loaded rock material are so small that they may be neglected in setting up equilibrium conditions.

These assumptions are supported by measurements of the observed strains within the elastic range in metals and rocks, with some exceptions of porous solids and other materials, to a sufficient degree for most practical applications.

In certain rock mechanics analyses, one thus usually adheres to the fundamental theory of elasticity, to the basic elasticity constants such as YOUNG's modulus of elasticity E, and POISSON's ratio $\mu = 1/m$, where $m = 1/\mu$ is the POISSON's number.

4-16. Plasticity

Plasticity of a solid material is its property to be continuously and permanently deformed, that is, a property to change shape in any direction without rupture under a stress exceeding the yield value of the material. Thus, plasticity of a material is characterized by the existence of a yield point beyond which permanent strains appear. In other words, the plastic deformation of a material is the permanent deformation after complete unloading of the material, assuming the unloading to be elastic. In the plastic state, permanent deformation of a material may occur without fracture. The term "fracture" implies the appearance of distinct surfaces of separation in the material.

The exact definition of plastic deformation, valid in all cases, refers to strain rates, of course. Plastic flow of a solid is the phenomenon of increasing strain at constant stress.

Physically, plastic deformation is anisotropic. Any initial isotropy which may have been present is usually destroyed by plastic deformation.

The conditions prevailing in the deeper strata of rock, such as long duration of small differences in principal stresses, elevated temperatures, and high average pressure, are all contributing to plastic deformations of rocks.

As shown in Chapter 8, plastic flow tends to relieve the high stress concentrations that would ordinarily develop in perfectly elastic materials.

Today, the plastic state of matter is of considerable interest to many branches of science and of engineering, among them rock mechanics and soil mechanics. The changes in minerals and rocks brought about by plastic deformation are in many respects analogous to certain phenomena observed in the changes in the structure of metals.

In studies of the plastic state of stress in soils and rocks, use is made of the so-called slip surfaces (refer to Sections 8-6 to 8-8). The slip surfaces are curved according to the general mathematical, viz., physical, curve of a logarithmic spiral (50-52), namely:

$$r = r_{\mathrm{i}} \cdot e^{\pm \omega \cdot \tan \varphi} \tag{4-36}$$

where r = general, variable radius vector;
r_{i} = initial radius vector;
e = base of the system of natural logarithms;
ω = amplitude (angle between r_{i} = const and variable r);
φ = angle of internal friction of soil or rock, whichever the case is;
$\tan \varphi$ = coefficient of friction.

Here r and ω are polar coordinates.

Equation (4-36) represents a family of two orthogonal, equilateral logarithmic spirals, viz., slip lines (Fig. 8-7).

Criteria of Rock Failure

4-17. General Notes about Failure Criterion

An important problem in rock mechanics is to ascertain the mechanical conditions which cause rock to deform permanently, or to fracture. If the mechanical, viz., stress condition in a rock is known, there arises immediately the question about the criterion of rock failure or fracture (rupture) and/or lasting plastic deformation without increase in stress. The term "fracture" is here used in the sense of brittle fracture or failure. This implies a complete loss of cohesion across the surface of failure.

The criteria of rock failure which are most useful in rock mechanics have not been derived merely from simple mathematical assumptions. Rather, the criteria are the expressions of physical hypotheses. Obviously, in designing in rock, and remembering that all physical laws have been derived from experiment, it is desirable to have a criterion of rock failure based on experiment. Such a theory should specifiy how the strength of the rock is affected by the state of stress in rock, time effects, temperature, and other factors involved. Unfortunately, no complete, plausible theory of failure has been put to the fore for a complex polycrystalline material, such as rock, for example. As a result, during the course of time, various empirical assumptions about the criteria for failure have been made.

In appraising the danger of a failure occurring in rock, a distinction should be made among some of the most commonly used failure criteria. Some of the better known criteria of failure that are based on experiment and are considered to be reasonably acceptable in soil mechanics as well as in rock mechanics are (25, 36, 36a, 64, 90):

1. The maximum tensile stress;
2. TRESCA's criterion, or that of the maximum shear stress;
3. COULOMB's criterion of failure;
4. MOHR's criterion; and
5. GRIFFITH's criterion of brittle failure in tension.

Because of their different physical background, it is well, here, to keep COULOMB's and MOHR's criteria separate.

Here, the term criterion is now advisedly used, because in a polycrystalline rock material where very little is known about the mechanisms of deformation and failure, it is impossible to devise a theory which will comprehend and fit all rocks in all states of stress, under various confinement, at various temperatures, and at all times.

4-18. Various Criteria

Maximum Tensile Stress Criterion for Rock

By this criterion, the rock material is assumed to fail by brittle fracture in tension when the applied least principal stress $-\sigma_3$ to the rock is equal to its uniaxial tensile strength $\sigma_{t\ ult}$:

$$\sigma_3 = -\sigma_{t_{ult}} \tag{4-37}$$

Maximum Shear Stress Criterion

TRESCA's criterion of failure is valid for isotropic, ductile materials. The criterion may be stated as a function of principal stresses σ_1 and σ_3.

According to this criterion, the material is assumed to fail when the maximum shear stress τ_{max} is equal to the shear strength s of the material (90):

$$s = \tau_{max} = \frac{\sigma_1 - \sigma_3}{2} \tag{4-38}$$

where σ_1 and σ_3 are the major and minor principal stresses, respectively. The intermediate principal stress σ_2 plays no part in this criterion. Hence, the decisive factor in TRESCA's criterion for failure is the maximum shear stress in the material.

One notices from Eq. (4-38) that TRESCA's failure criterion is a special case of COULOMB's criterion (refer to Section that follows).

Coulomb's Criterion of Failure

Empirically, the shear strength s of a cohesive soil or rock varies with the normal stress σ_n on the rupture plane according to the classical COULOMB's law written as (25):

$$s = \tau = \sigma_n \cdot \tan\varphi + c \tag{4-39}$$

where τ = shear stress,
φ = shearing strength parameter or angle of friction (a test parameter), and
c = test parameter known as the cohesion.

Geometrically, Eq. (4-39) represents a straight line, *t-t* (Fig. 4-4), known as the COULOMB's shear strength line.

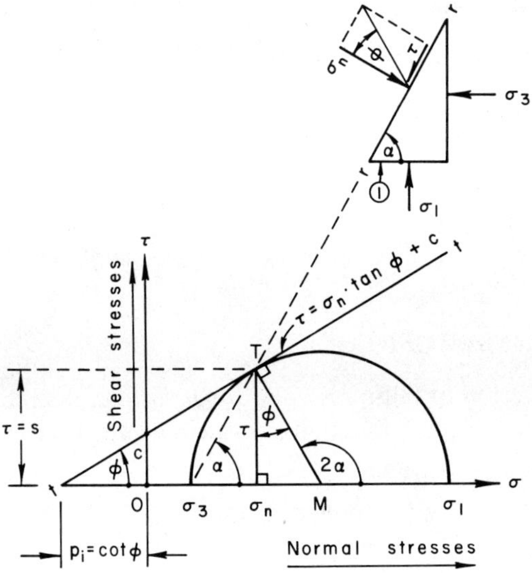

Fig. 4-4
Coulomb-Mohr failure criterion (general case)
r – r: rupture or failure plane
t – t: Coulomb's shear strength line, tangent to circle
$\sigma_1 - \sigma_3$: diameter of Mohr's stress circle
$\tau = s$: shear strength

From geometry as in Fig. 4-4, the normal stress σ_n on the rupture or shear plane *r-r* is calculated as

$$\sigma_n = \frac{\sigma_1 + \sigma_3}{2} + \frac{\sigma_1 - \sigma_3}{2} \cos 2\alpha \qquad (4\text{-}40)$$

wherein σ_1 and σ_3 are the major and minor principal stresses, respectively, and α is the angle of rupture.

The shear stress τ in the rupture plane is:

$$\tau = \frac{\sigma_1 - \sigma_3}{2} \cdot \sin 2\alpha \qquad (4\text{-}41)$$

When $c = 0$, the shear strength at failure is, according to Eq. (4-39):

$$s = \tau = \sigma_n \cdot \tan \varphi \qquad (4\text{-}42)$$

a straight line passing through the origin of the normal stress/shear stress coordinates (Fig. 4-5).

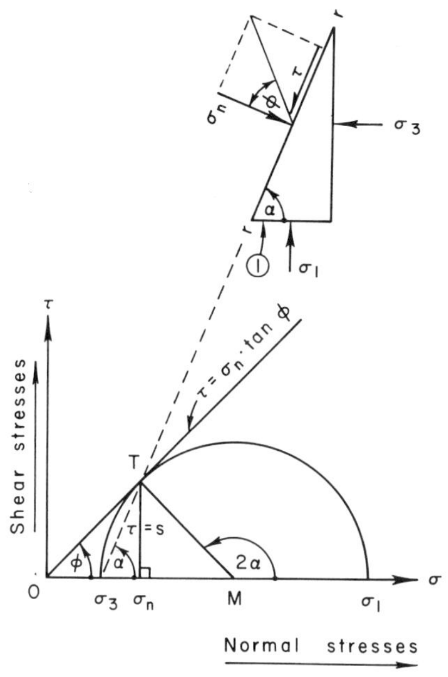

Fig. 4-5
COULOMB-MOHR failure criterion when $c = 0$

When $\varphi = 0$, then Eq. (4-39) transforms into:

$$s = \tau = c = \frac{\sigma_1 - \sigma_3}{2} = \text{const} \tag{4-43}$$

(Fig. 4-6), i.e., the shear strength is constant and independent of normal stress. One sees that TRESCA's criterion is a special case of COULOMB's criterion for failure.

Furthermore, when $\sigma_3 = 0$ and $\varphi = 0$, then:

$$s = \tau = c = \frac{\sigma_1}{2} \tag{4-44}$$

i.e., the shear strength of a pure cohesive material is equal to one-half its compressive strength σ_1.

Accepting MOHR's criterion, the discussion of which follows immediately in the next section, one notices in Fig. 4-4 that COULOMB's criterion is equivalent to MOHR's linear envelope to MOHR's stress circles.

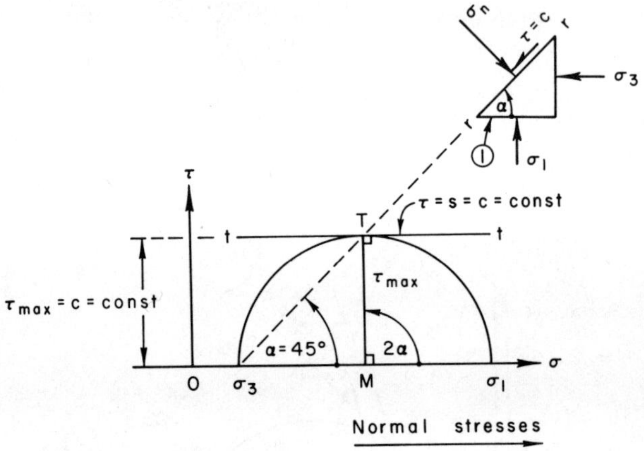

Fig. 4-6
COULOMB-MOHR-TRESCA failure criterion for a pure cohesive material ($\varphi = 0$)

COULOMB's shear strength criterion may also be written from geometry as in Fig. 4-4 in terms of principal stresses as:

$$\sin \varphi = \frac{\sigma_1 - \sigma_3}{\sigma_1 + \sigma_3 + 2\, p_i} \tag{4-45}$$

where

$$p_i = c \cdot \cot \varphi \tag{4-46}$$

is the initial stress of the test specimen.

When $c = 0$, $p_i = 0$, and

$$\sin \varphi = \frac{\sigma_1 - \sigma_3}{\sigma_1 + \sigma_3} \tag{4-47}$$

Referring to Fig. 4-4, it can be reasoned out that as the confining pressure σ_3 is increased, the normal stress σ_n also increases on the incipient plane of shear (rupture). Thus, the necessary shear stress τ (s) to bring about failure also increases. From geometry, as in Fig. 4-4 or Fig. 4-5:

$$2\alpha = 90° + \varphi \tag{4-48}$$

and

$$\alpha = 45° + \varphi/2 \tag{4-49}$$

meaning that, theoretically, failure in shear takes place on rupture planes at an angle of rupture $\alpha = 45° + \varphi/2$ to the major principal plane ①.

Theoretically, in the case of a pure cohesive material, Fig. 4-6 shows that the rupture plane intersects the line of action of the axial load at $\alpha = 45°$. In reality, however, the angle of rupture α varies from one type of rock to another.

Although COULOMB's straight-line criterion is used extensively to predict failure in rocks, it must be said that this criterion does not represent exactly the curvilinear envelope to MOHR's stress circles of most rocks.

Mohr's Criterion of Failure

MOHR's criterion merely assumes the existence of a shear failure envelope (Fig. 4-7), which itself may be a straight line (*t-t*), or a curvilinear one (*e-e*). The envelope in MOHR's criterion is the governing condition for the occurrence of rupture surface brought about by the shear stress upon exhaustion of the shear strength of the material. Analogous to soil mechanics, MOHR's failure criterion can also be used in rock mechanics.

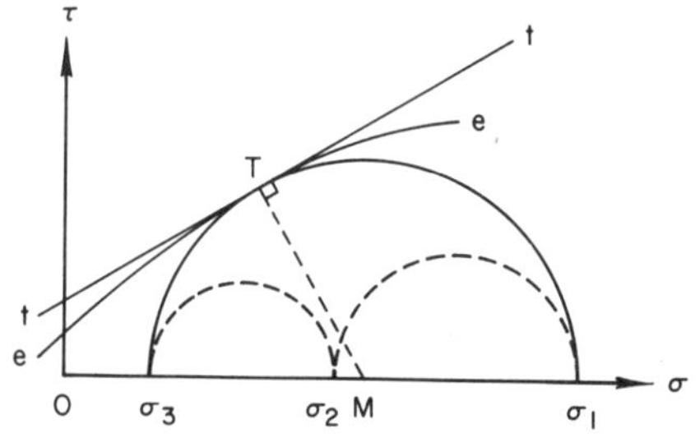

Fig. 4-7
Straight-line (*t* — *t*) and curvilinear envelopes in MOHR's criterion of failure

If the stress-circle envelope is a straight line, then MOHR's and COULOMB's criteria are identical (64).

MOHR's failure criterion says that a compressed material fractures when the shear stress τ on the incipient rupture plane increases to a value which depends upon the magnitude of the normal stress σ_n on the rupture (shear) plane, i.e., $\tau = f(\sigma_n)$. This relationship has been established experimentally.

On MOHR's stress circle diagram as in Fig. 4-7, it can be seen that failure of rock (soil) takes place when the $(\sigma_1 - \sigma_3)$-diameter stress circle just tangents the shear strength envelope *e-e*. Also, one notices in this diagram that in the three-dimensional representation of MOHR's failure criterion, the intermediate principal stress σ_2 (if any) does not affect failure. Only the major (σ_1) and minor (σ_3) principal stresses play an active part in this criterion. One should also be cognizant that in MOHR's stress-circle diagram, the failure envelope (*e-e*) is not represented by any mathematical formula but is obtained experimentally: the envelope is drawn to tangent the various stress circles of a set of experiments (Fig. 4-8).

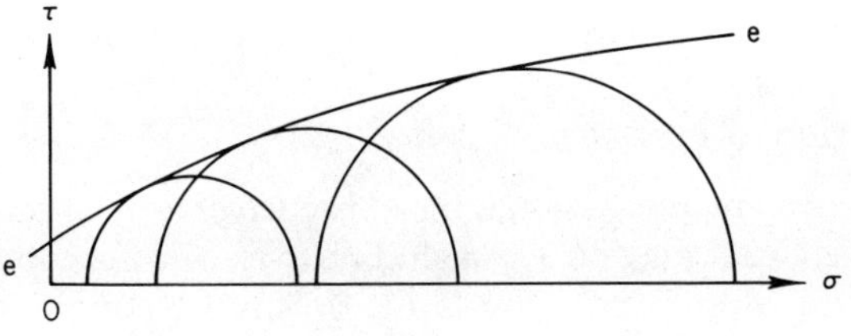

Fig. 4-8
MOHR's curvilinear strength envelope $e - e$

MOHR's failure criterion specifies on a macroscopic scale not only the state of stress at failure, but also the direction of the failure plane. However, COULOMB's and MOHR's criteria do not account for the cause of failure of the rock material on an internal or microscopic basis.

Although MOHR's criterion does not explain completely and satisfactorily all the observed variations and deviations in rock compressive and shear strength, at the present time it enjoys its widest acceptance in rock mechanics.

MOHR's failure criterion can also be used with reasonable validity in studies of shear strength in rock faults, joints, and other kinds of discontinuities (Fig. 4-9).

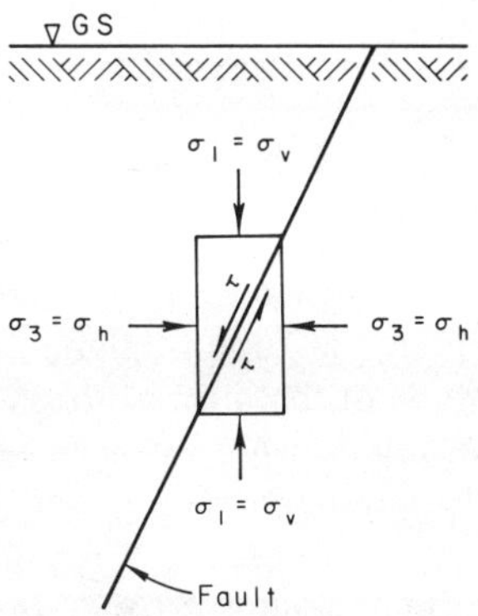

Fig. 4-9
Shear in a fault

Griffith's Criterion of Tensile Failure

It is known that crystalline substances contain microfractures. In his criterion of failure, GRIFFITH (36, 36a) assumes the existence of thin, flat, narrow, elliptical uniform microcracks in the material. These cracks bring about stress concentration at the tips or ends of GRIFFITH's microcracks. According to GRIFFITH's criterion, fracture of material is caused by these stress concentrations, causing the crack to propagate and ultimately contributing to macroscopic failure of the material. Also, the GRIFFITH criterion postulates that fracture is initiated in a brittle material by failure in tension around the tips of the microcracks, and that the crack retains its form until the moment of failure. GRIFFITH's hypothèsis has been corroborated by experimental work on glass. Insufficient studies have been made on rocks to determine whether GRIFFITH's criterion would predict compression failure by knowing the tensile strength of the rock.

As to rock, GRIFFITH's theory does not account for the closure of the crack before failure. If the cracks do close, it can be expected that frictional forces will be acting between the closed surfaces. In such a case, GRIFFITH's criterion becomes a special case of COULOMB's criterion.

The actual mechanism of failure in rock is still a subject of hypothesis. Despite its theoretical elegance, there can be little general application of GRIFFITH's failure criterion to rocks as yet, mainly because the varying rock properties require an empirical approach as represented by the COULOMB or MOHR failure criteria.

Also, joints in rock are not equivalent to thin elliptical openings, because joints in situ more commonly abut a network of other joints. Besides, the lengths of the cracks in a rock mass are not small with respect to the stressed rock volume, and the spacings of the joints are frequently close enough to affect the stress concentrations around any flaw. The stress concentrations around a GRIFFITH elliptical crack are computed based on the theory of elasticity. Therefore, the mechanism of failure should be time-independent, and hence would not account for variation in strength with stress or strain rate.

In the words of FARMER (32), "the GRIFFITH's criterion applied to rock is at best a rather complicated way of arriving at a logical solution."

Deformation of Rocks

4-19. Definition

The manner in which a rock or rock mass deforms upon action of loads on them, or because of changes in internal stresses brought about by all kinds of excavations in rock, is an important consideration in rock mechanics.

The term "deformation of rock" means any change in the original form or volume of a rock specimen; or change by externally applied loads on in-situ rock; or by tectonic forces (compressive and/or shear forces). In nature, the common modes of deformation may be folding, faulting (shear), and solid flow.

4-20. Mutual Dependence of Stress and Strain

In studies of rock deformations and rock strength, as well as in theory and design of structures in rock, the concepts of stress and strain are basic: every stress brings about strain and displaces individual rock particles. The concepts of stress and strain are also important in studying stress fields.

Upon subjecting a test specimen of rock to compression, it deforms: to a certain particular stress there occurs a corresponding definite deformation, viz., strain. The rock test specimen in the testing machine resists the deformation. The specimen props firmly against the externally applied load with an induced stress that corresponds to the externally applied load. Thus, the simultaneously induced stress in the rock and its deformation depend upon each other.

Without deformation there is no stress possible, and without stress no deformation can be brought about. Also, every stress imparting a strain on the rock displaces individual rock particles.

4-21. Stress-Strain Diagrams

For any axial symmetrical load applied to the cylindrical rock core specimen, the axial and lateral deformations are measured and the corresponding strains ε_1 and ε_3 are calculated. The plot of stresses and corresponding strains of the tested rock results in a stress-strain diagram.

Now, in the discussion that follows, and for the sake of one's indoctrination in the understanding of the stress-strain relationships of various materials, let us avail ourselves to the study of several stress-strain diagrams that follow.

In Figs. 4-10a, b, and c are shown three idealized materials characterized by the modes according to which they deform. Figure 4-10a illustrates elastic deformation of an elastic material represented by a straight-line stress-strain (force-deformation) diagram. Figure 4-10b shows viscous deformation of a viscous substance represented by an straight-line shear-force versus rate-of-shear diagram. Figure 4-10c represents plastic deformation of a perfectly plastic substance represented by a straight-line stress-strain diagram.

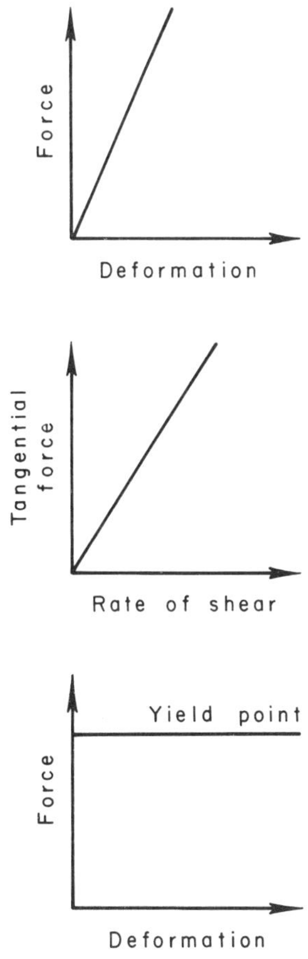

a) Elastic substance represented by a straight line. Elastic deformation

b) Viscous substance
Viscous deformation

c) Perfectly plastic substance. Plastic deformation or plastic flow condition

Fig. 4-10
Three idealized substances a), b) and c), characterized somewhat by the modes according to which they deform

Factors Affecting Deformation

Rock deformation and strength are affected by the so-called deformability factors. These are the geological details such as:

1. Various rock defects;
2. Rock petrographic structure (rock matrix);
3. Geometric orientation and attitude of the rock formation (dip and strike);
4. Degree of weathering or alteration of the rock;
5. Elastic, plastic, and rheological properties of rock;
6. Anisotropy of rock;
7. Direction and magnitude of acting loads on rock;
8. Degree of compression and/or decompression of rock;
9. Fissures and/or hair cracks brought about by blasting and/or excavation of, or drilling in, the rock;
10. Seismic factors; and
11. State of stress (internal stress) within the rock mass.

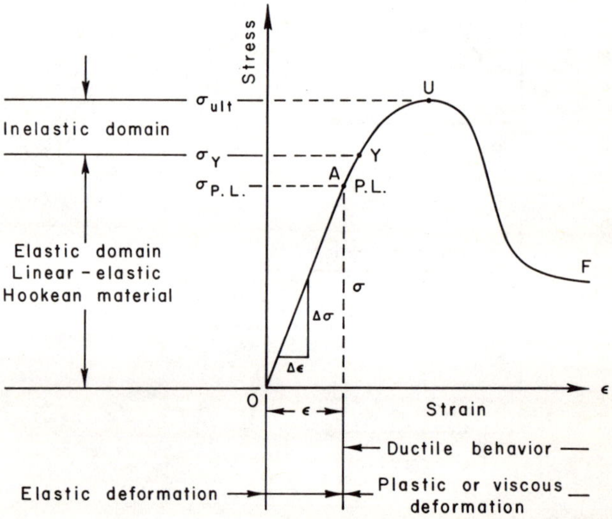

Fig. 4-11
Stress-strain diagram

Hence, because of the anisotropy of the rock, rock deformations cannot be characterized by a single parameter. Also, deformations depend upon the kind of load imposed on rocks—i.e., whether static, dynamic, or cyclic—and their magnitude and duration (creep, for example). Besides, plasticity, too, affects deformability and magnitude of deformations of rocks considerably.

Whereas elastic deformations entail changes in the volume of rock, plastic deformations (viz., the relative movement or sliding of the ruptured or sheared-off parts of the rock past each other) take place usually at a constant volume.

Slow deformation of rock at a constant load (solid flow) is a rheological phenomenon known by the term "creep." It is a feature of rock like creep in concrete. Creep contributes to genuine pressure in a rock mass.

Figure 4-11 shows an idealized stress-strain relationship diagram for a ductile material. In this diagram, it can be observed that up to a certain magnitude of the externally applied stress σ_Y (yield point Y) on the rock, the corresponding strain ε is constant. In other words, the stress σ (or $\Delta\sigma$) is proportional to strain ε (or $\Delta\varepsilon$) by HOOKE's law *"ut tensio sic vis"*:

$$\sigma = E \cdot \varepsilon \tag{4-50}$$

or

$$\Delta\sigma = E \cdot \Delta\varepsilon \tag{4-50a}$$

where $E = \sigma/\varepsilon$ = HOOKE's coefficient of proportionality, known also as YOUNG's modulus of elasticity, defined for uniaxial stress; the modulus of elasticity E represents the stiffness of the material.

When $\varepsilon = \Delta L/L = \sigma/E = 1.0$, then $E = \sigma$. Here L = original, axial length of test specimen, and ΔL = absolute linear axial deformation of the same test specimen.

It should be recalled that for most materials, the proportional limit (point *P.L.*) practically coincides with the yield point (point Y). For this reason and for the sake of brevity, only the terms yield stress σ_Y and yield point Y will be used in the discussion that follows.

The other elasticity constant of the elastic material used in its strength analysis is POISSON's ratio μ:

$$\mu = 1/\text{m} = \varepsilon_3/\varepsilon_1 \tag{4-51}$$

where $m = 1/\mu$ is POISSON's number, ε_3 is the lateral or transverse strain, and ε_1 is the longitudinal strain of the material tested.

The so-called elasticity constants E and μ may be obtained experimentally not merely from uniaxial stress-strain diagrams, but also from biaxial and triaxial stress-strain diagrams as well.

Referring to HOOKE's law Eq. (4-50), one notices that between the stress and strain a linear relationships exists. This means that up to point Y in Fig. 4-11, the strains are elastic strains, and the material is said to be linear-elastic. A material is called linear-elastic if the relationship $\sigma = E \cdot \varepsilon$ holds accurately.

It should be kept in mind that the E-value is different for various materials.

Point Y, at which the transition from elastic to ductile behaviour of the material takes place, is termed the *yield point,* and the corresponding stress σ_Y is called the *yield stress.* Plasticity is characterized by the existence of a yield point beyond which permanent strains appear.

Upon continuing the loading beyond the yield point Y up to the ultimate stress σ_{ult} to point U on the stress-strain curve, failure of the material takes place, and with increasing strain beyond U the stress drops, as shown by the part UF of the curve. In other words, the material deforms without limit under this stress unless constrained. Thus, Figure 4-11 represents a stress-strain curve for a perfectly elasto-plastic material.

In this stress-strain diagram, the following three domains can be distinguished, namely:

1. The elastic domain,
2. The partly elastic-plastic domain, and
3. The plastic domain,

the domains as in 2 and 3 above being called the *inelastic* domain. Deformation in the inelastic domain is generally termed plastic or viscous deformation.

A material is called *purely elastic* if the deformation is recovered when the stress is removed.

A material is called *purely plastic* if the deformation does not disappear when the stress is removed and if the stress determines the amount, but not the rate, of deformation.

A material is called *purely viscous* if the stress acting upon it determines the rate of deformation.

Theoretically, as mentioned already earlier in this Chapter, within the domain of elasticity of an ideal elastic material, unloading or removing of stress recovers the corresponding strain completely. That is, there should be no permanent, irreversible deformation in the material.

Refer now to Fig. 4-12. If upon loading beyond the yield point Y the stress-strain curve continues to rise within the inelastic domain from Y to F above yield point, the material is said to be *strain-hardened*.

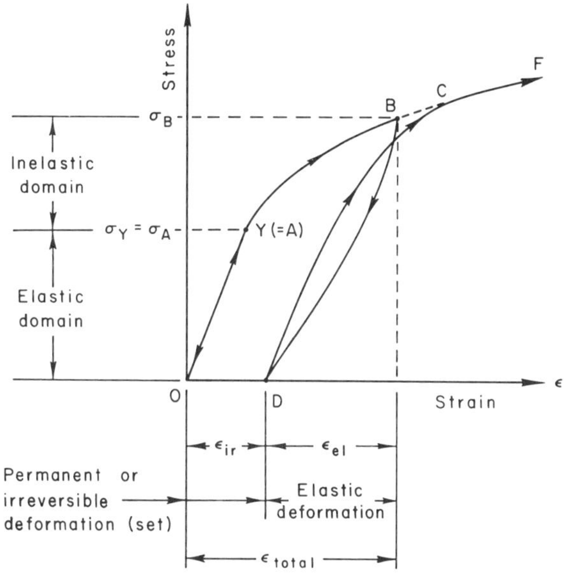

Fig. 4-12
Stress-strain diagram of strain hardening characteristic of a ductile material

If the material is unloaded from σ_B to $\sigma_D = 0$, then only part of the strain is recovered: an *irreversible,* permanent, relative deformation or strain (permanent set) of the magnitude ε_{ir} results (Fig. 4-12), known also by the term plasticity deformation:

$$\varepsilon_{ir} = \varepsilon_{total} - \varepsilon_{el} \tag{4-52}$$

Upon reloading the material with the stress σ_B, the stress-strain curve takes the course CF. Thus it can be seen that strain-hardening increases the yield stress from σ_A to σ_B but does not affect the elastic constant of the material.

The ratio e_{el} of the elastic strain ε_{el} to the total strain ε_{total} of the material is termed the *degree of elasticity* of the rock:

$$e_{el} = \varepsilon_{el}/\varepsilon_{total} \tag{4-53}$$

The ratio e_{pl} of the plasticity strain $\varepsilon_{pl} = \varepsilon_{ir}$ to the total strain ε_{total} of the material is termed the *degree of plasticity* of the material:

$$e_{pl} = \varepsilon_{ir}/\varepsilon_{total} \tag{4-54}$$

Unloading of rock in underground structures and rock slopes is an important stress release phenomenon around every underground opening, every hole, and every open cut. This phenomenon is based on plastic performance of the rock, and lies within the plastic domain of the stress-strain diagram.

In repeated loading-unloading-loading tests, the ratio E_c of stress σ to the total strain $\varepsilon_{total} = \varepsilon_{el} + \varepsilon_{ir}$ is called the *compression modulus* or *modulus of deformation:*

$$E_c = \frac{\sigma}{\varepsilon_{total}} = \frac{\sigma}{\varepsilon_{el} + \varepsilon_{ir}} \tag{4-55}$$

This modulus is thus based on the total measured strains, i.e., elastic plus inelastic (irreversible or plastic) strains, ε_{el} and ε_{ir}, respectively.

The true modulus of elasticity E from such loading test results is then expressed on the elastic strain basis as:

$$E = \frac{\sigma}{\varepsilon_{total} - \varepsilon_{ir}} \tag{4-56}$$

Depending upon the amount of jointing in a rock mass, viz., elastic and plastic strains, the deformation modulus E_c may be quite different from the true, elastic modulus of elasticity E.

4-22. Stress-Strain Diagrams for Rocks

As to the deformations of rock, it has been learned that usually many kinds of rock fail, or fracture at the proportional limit P.L. of elasticity or somewhat beyond it very near to the yield stress. Such materials are referred to as brittle. Rocks are characterized as failing in brittle fracture if they fail with no previous plastic deformation. Rocks are referred to as ductile if they deform appreciably (plastically) before failure. After previous plastic deformation, the rock fails by ductile rupture. For ductile materials, there is no brittle fracture.

Under normal temperatures and pressures, rocks usually tend to exhibit a brittle kind of rupture of failure mechanism.

The term "fracture" is used here in the sense of brittle fracture or failure; this implies a complete loss of cohesion across a surface.

Because most of the rock materials are brittle, the plasticity domain of such rocks and thus their plastic deformation and degree of plasticity are relatively very small.

For most rocks, the stress-strain diagrams take an approximately linear course, like that of a perfectly elastic solid where stress is proportional to strain and where there is no yield point, ending abruptly in failure (at point F on the diagram as in Fig. 4-13a). If the relationship $\sigma = E \cdot \varepsilon$ for such a material holds strictly, the material is referred to as linear-elastic. Generally, igneous rocks and sedimentary rocks under ordinary compression loads deform very little before they break. However, at high temperatures and/or pressures, rocks deform plastically. References 38 and 39 contain many papers dealing with such experimental data.

The brittle-ductile transition is of great interest in geology and geophysics in connection with the behavior of rock materials in the lower earth's crust. It is not of great interest in rock mechanics, because the necessary large pressures and high temperatures rarely are encountered in engineering practice.

In Figure 4-13b is shown a σ-ε-diagram for what is termed by FARMER (32) a *quasi-elastic* rock. These rocks approximate a brittle elastic material with a nearly linear σ-ε-relationship to the point of failure. Examples of such rocks are the sound, dense, fine-grained, extrusive igneous rocks and fine-grained metamorphic rocks.

Figure 4-13c shows a σ-ε-diagram in which the slope decreases as the stress increases. Termed by FARMER *semi-elastic*, these rocks represent the less elastic, coarser-grained igneous rocks and fine-grained, densified, low-porosity sediments with a reasonable amount of cohesion.

Figure 4-13d illustrates a stress-strain diagram for a perfectly elastic material. Because of its nature, the diagram shows three kinds of modulus of elasticity, namely:

1. E_i — the initial tangent modulus at zero load
2. E_t — the tangent modulus at a particular point T on the stress-strain diagram for a specified stress (also shown is the 50% tangent at point P), and
3. E_s — the secant modulus for a particular point T.

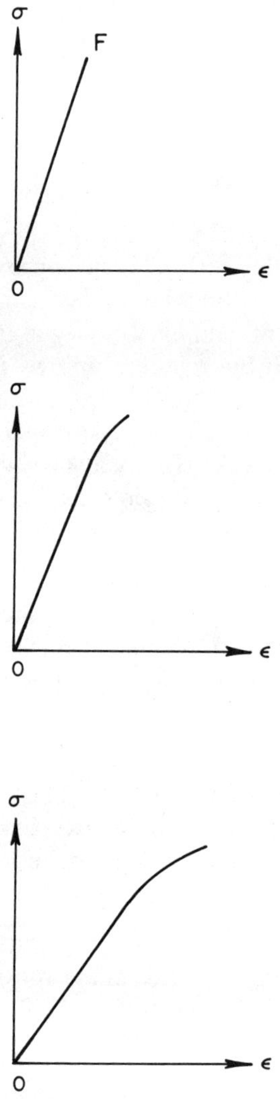

a) Linear-elastic rock

b) Quasi-elastic rock

c) Semi-elastic rock

Fig. 4-13
Various kinds of stress-strain diagrams for rocks

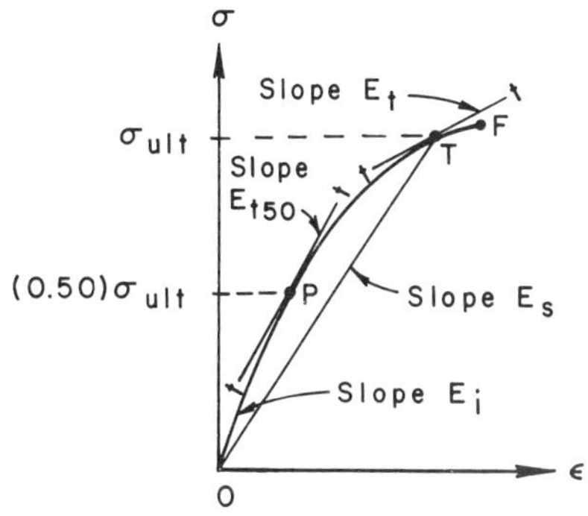

d) Generalized stress-strain diagram for rock. Perfectly elastic material showing 50 % — tangent modulus E_{t50}, and secant modulus E_s (OT)

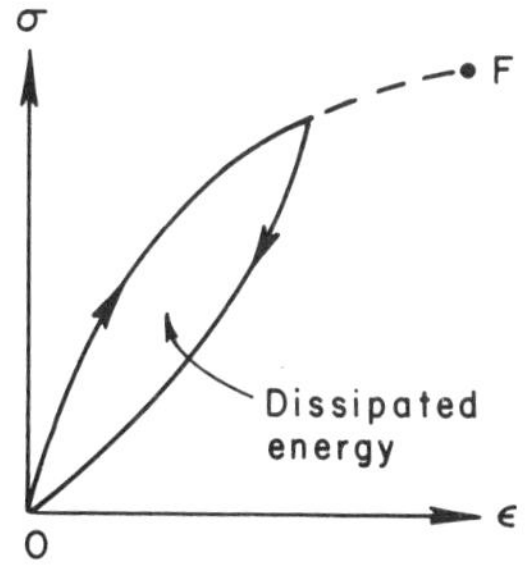

e) Elastic material with a hysteresis loop formed upon a single loading and unloading cycle

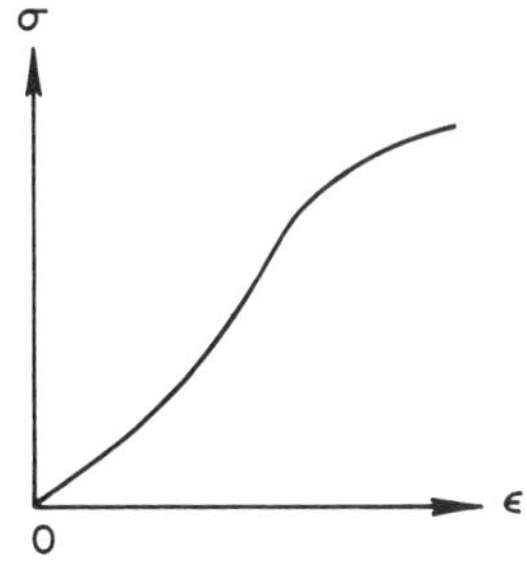

f) Nonelastic material

Fig. 4-13
Various kinds of stress-strain diagrams for rocks

For rock, normally the initial tangent modulus E_i is referred to because it is the most accurately obtained under test conditions.

Figure 4-13e shows a stress-strain diagram for an elastic material with a hysteresis loop formed by a single loading and unloading cycle.

Figure 4-13f depicts a variable stress-strain diagram for nonelastic material such as a less cohesive and weak sedimentary rock with large void spaces.

A different kind of testing methodology applied to the same kind of rock will result in different stress-strain diagrams, of course.

Creep

No material is perfect. All materials combine in some proportion the characteristics of elasticity, plasticity, and flow.

Plastic flow of rock is of particular interest in rock engineering in the design of underground openings at great depth below ground surface. When a material deforms slowly in a continuous way, this kind of permanent deformation is known as creep—a deformation phenomenon when strain increases linearly with time.

Creep—a complex response of strain to stress—is the time-dependent movement (plastic displacement) of rock under a sustained load. In other words, creep is the phenomenon of increase in strain during the course of time under constant stress. The amount of creep depends upon the stress level: at a high stress, creep accelerates, and failure is attained quickly.

It appears that no general equation exists that adequately defines the rheological properties of rocks. The various rheological formulas in the technical literature do not give any consideration for accelerated strain. For the same reason, the behavior of most of the real rock materials cannot be represented by rheological models. However, rheological models in rock mechanics increase our awareness of the variety of strain-time patterns that might exist in rocks.

Probably the most accurate creep data on rock to be used in design are obtained experimentally.

Figure 4-14 illustrates elastic and creep strains ε at constant stress σ as a function of time. If strain increases linearly with time, the deformation is known as creep. Here, in this experiment, essentially a constant stress (σ = const) is applied to a rock specimen, and the corresponding strain ε is a function of time; i.e., $\varepsilon = f(t)$.

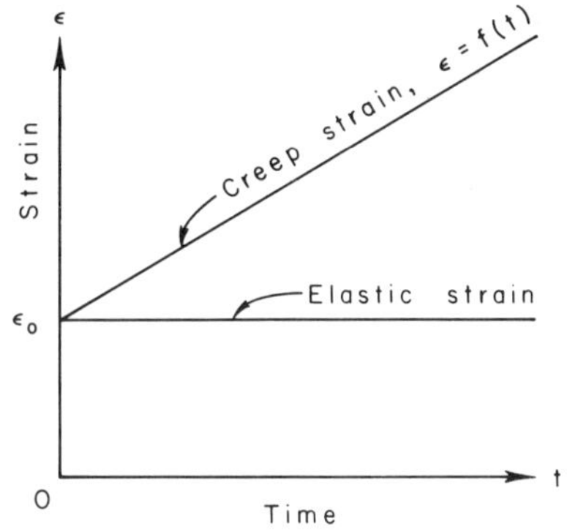

Fig. 4-14
Elastic and creep strains at constant stress

The creep test, generally, is one of more qualitative than quantitative nature, because it describes behavioral characteristics of rock rather than furnishes numerical values.

Figure 4-15 illustrates time-dependent deformation at constant stress. In time-dependent deformation such as in rock creep, three major phases

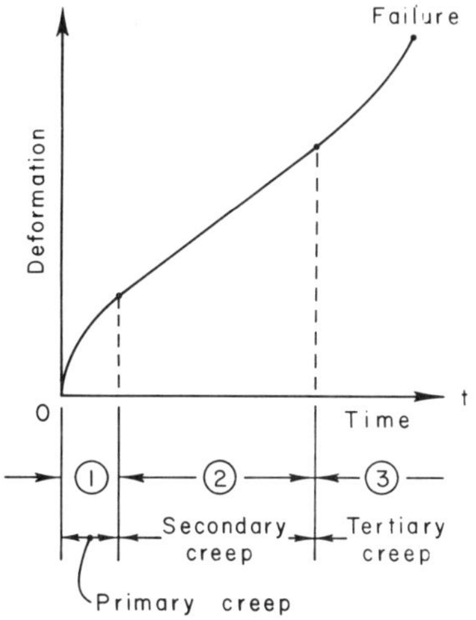

Fig. 4-15
Time-dependent deformation at constant stress

of strain responses to a constant stress can be visualized, namely:

1. Primary or transient creep,
2. Secondary or steady-state creep (pseudo-viscous flow) at a medium high stress), and
3. Tertiary or accelerated phase of creep—the phase of failure—occurring at high pressure.

In the primary phase, strain increases at a decreasing rate. In the second phase, the rate of strain is constant. During the tertiary phase of creep, strain increases at an increasing rate until the specimen fails or ruptures. This accelerated rate of strain in the tertiary phase occurs when the specimen becomes weaker and increasingly fractured.

4-23. Appearance of Deformed Rock Specimens

Some typical appearances of deformed rock specimens of brittle fracture to ductile flow by compression are shown schematically after Griggs and Handin (1960) in Fig. 4-16.

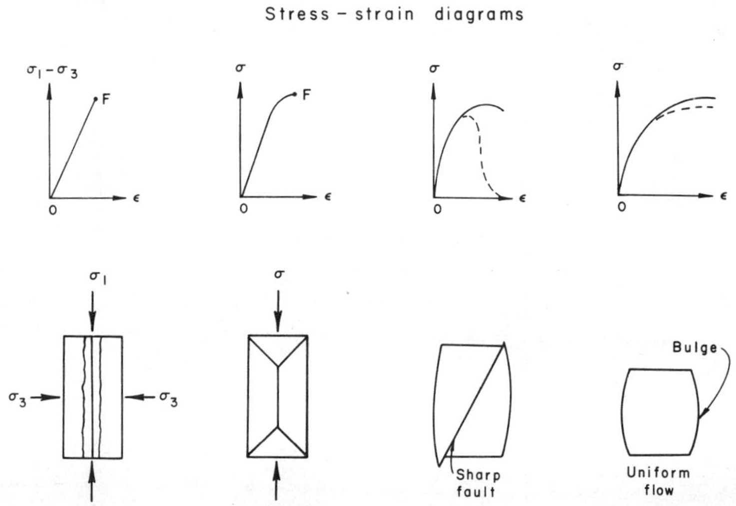

Triaxial compression. Failure in tension | Uniaxial compression. Failure in tension and shear | Failure in shear | Plastic failure

Fig. 4-16
Typical appearance of deformed rock specimens of brittle fracture to ductile flow

KÁRMAN's tests (58) on marbles and sandstones have shown, though, that the same material may be brittle or plastic, depending upon the magnitude of the confining pressure. Thus, testing conditions such as, for example, increase in either confining pressure σ_3 or in temperature, increase the ductility of the rock specimen and bring about the material to transform toward uniform flow (70).

Because of the varied appearance of the stress-strain diagrams, it is impossible to conclude from such diagrams alone the nature of such rock deformation.

Depending upon the constraint offered by the load-bearing platens of the compression testing machine, as well as upon the quality of the parallel end surfaces (smooth or rough), rock core specimens tested for their unconfined compression strength fail either in tension or in shear (Fig. 4-17). The greater the friction between the platens and the rock test specimen end surfaces, the more likely the specimen will fail in shear.

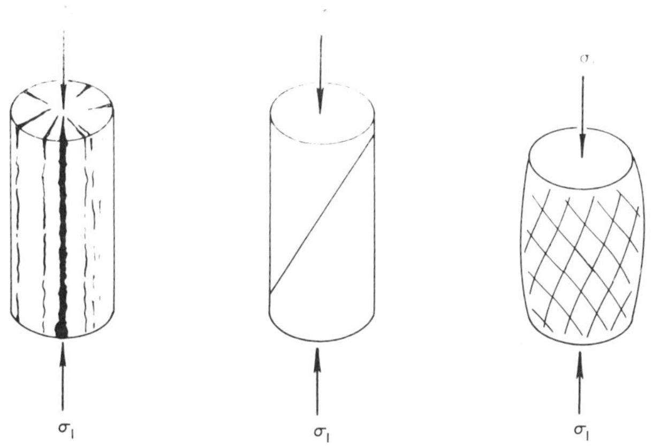

a) Smooth end surfaces b) Rough end surfaces c) Plastic deformation by bulging

Fig. 4-17
Some modes of failure of rock in unconfined compression

Fig. 4-17a shows an irregular, longitudinal splitting of a rock core specimen resulting from an unconfined compression test.

Fig. 4-17b shows a rock failure in shear. Its characteristic is shear displacement along the surface of rupture, viz., shear plane corresponding to a geological fault.

Fig. 4-17c depicts a multiple shear failure; the plastic deformation is accompanied by a network of shear planes.

Figure 4-18 shows the appearance of some deformed rock cores subjected to unconfined compression test. Specimen a)—a pegmatite gneiss—failed by axial splitting. Specimen b)—also a pegmatite gneiss—failed from the characteristic cones formed on the shear planes. Specimen c), a Vermont marble, and specimen d), a Triassic Brunswick shale, failed in shear.

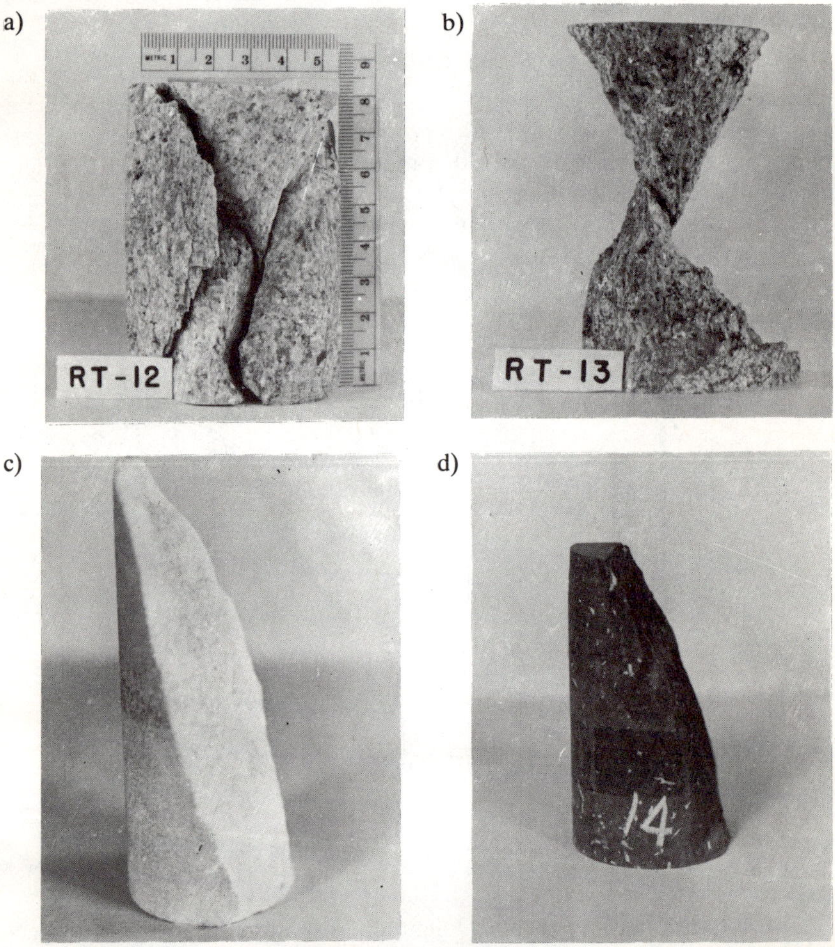

Fig. 4-18
Deformations of some rock core specimens:
a) and b) Pegmatite gneiss
c) Vermont marble
d) Triassic Brunswick shale

From the discussion on σ-ε relationships, it becomes apparent that generally rock deformation properties cannot be studied on the basis of merely simplifying assumptions, as is usually done for metals. Nor can rock deformations be described by a single parameter. This is so because of heterogeneity and anisotropy of rocks, and the testing methodology used. Also, rock plasticity greatly affects rock deformation and stress reversibility; viz., full recovery of strain after unloading the rock is not ideally possible even in the elastic range of testing.

4-24. Modulus of Elasticity of Rocks

Knowledge about the modulus of elasticity of rocks is an important factor in evaluating rock deformation under various loading conditions. It is also necessary in studies of seismics in rocks. The modulus of elasticity of rock varies from one geologic region to another because of the existence of various types of rock formation. For this reason alone, different values of the modulus of elasticity of different rocks can be anticipated. Hence, there exists a need for the determination of the numerical values of the modulus of elasticity of rocks.

Generally, the modulus of elasticity of rock is affected by rock type, porosity, particle size, and water content. The variation in the values of the modulus of elasticity is more pronounced in clayey shales than in sandstone, for example.

The modulus of elasticity is greater perpendicular to stratification and fissure than that determined parallel to the bedding planes and fissures.

The various modes of genesis or formation of rocks also affect the magnitude of the modulus of elasticity.

The reason for high modulus of elasticity values of some rocks probably lies in their structures themselves, as well as in their mineral chemical compounds. The known values of the modulus of elasticity of various rocks may suggest that eruptive igneous rocks of basic nature—like basalts, for example —have a larger modulus value than acidic rocks—such as granite, for example.

Technically, the modulus of elasticity of a rock can be increased considerably by means of rock grouting.

Elasticity constants E and μ of rocks are determined by:

1. Static tests, and
2. Dynamic tests, in which velocities of propagated energy waves are measured. Measurements of the shear wave (S-wave) velocity thus affords the determination of POISSON's ratio of geologic materials.

TABLE 4-9

Elasticity Constants E and μ of Some Rocks

Rock	YOUNG's modulus of elasticity E		POISSON's ratio	References
	kg_f/cm^2	N/m^2	μ	
	Multiply by 10^5	Multiply by 10^{10}	Average values	
1	2	3	4	5
Igneous rocks				
Basalt	2.0–10.0	1.96– 9.81	0.14 –0.25	32, 54, 55, 86
		4.85–11.15	0.22 –0.25	18
Diabase	3.0– 9.0	2.94– 8.83	0.125–0.25	34, 54
		–	0.333	55
		2.20–11.40	0.103–0.184	55
		8.00–10.75	–	18
Gabbro	6.0–11.0	5.88–10.78	0.125–0.25	32, 54, 55
		5.84– 8.71	0.154–0.48	18
Granite	2.6– 7.0	2.55– 6.86	0.125–0.25	32, 54
		–	0.155–0.338	18
		–	0.150–0.240	22, 55
		2.13– 7.05	–	18
Syenite	6.0– 8.0	5.88– 7.85	0.25	32
		6.29– 8.63	0.17 –0.319	18
			0.15 –0.34	55
Sedimentary rocks				
Dolomite	2.0– 8.4	1.96– 8.24	0.08 –0.20	32, 54
		7.10– 9.30	0.08 –0.20	54, 86
		–	0.32 –0.37	18
Limestone	1.0– 8.0	0.98– 7.85	0.10 –0.20	54, 86
		–	0.16 –0.23	55
		–	0.33 –	54
		0.80– 2.10	0.14 –0.30	18
Sandstone	0.5– 8.6	0.49– 8.43	0.066–0.125	22
		–	0.230–0.300	86
		–	0.17	54, 55
		–	0.07	22
		–	0.62	49
Shale (clay)	0.8– 3.0	0.78– 2.94	0.11 –0.54	32, 55
		0.98– 2.35	0.10	49
		1.20– 4.40	0.23 –0.30	86
		–	0.04 –0.12	18
Sandstone	4.5– 5.2	4.41– 5.10	0.21 –0.24	54

TABLE 4-9 (Continued)

Elasticity Constants E and μ of Some Rocks

Rock	YOUNG's modulus of elasticity E		POISSON's ratio	References
	kg_f/cm^2	N/m^2	μ	
	Multiply by 10^5	Multiply by 10^{10}	Average values	
1	2	3	4	5
Metamorphic rocks				
Gneiss	2.0– 6.0	1.96– 5.88	0.091–0.25	32, 54
	2.5– 6.0	2.45– 5.88	0.11	22, 86
		1.42– 7.00	0.03 –0.15	49
		–	0.09 –0.20	86
Marble	6.0– 9.0	5.88– 8.83	0.25 –0.38	22, 55
		– 8.50	0.25	
		4.93– 8.70	0.16 –0.27	18
		2.8 –10.00	0.11 –0.20	54
Quartzite	2.6–10.2	2.55– 8.70	0.23	22
		2.80– 8.70	0.11 –0.20	54
		9.75	0.15	18
Schist	4.1– 7.2	4.0 – 7.05	0.01 –0.20	18
		–	0.08 –0.20	22, 55
		–	0.10 –0.17	54
Slate			0.06 –0.44	

For the purpose of one's orientation, some values of the modulus of elasticity E and POISSON's ratio μ for various kinds of rocks as found in the technical literature are compiled in Table 4-9 (18, 22, 24, 32, 47, 49, 54, 55, 73, 83, 91, 92).

The variation in the values of the elasticity constants of rocks may be partly attributed to the nonhomogeneity and anisotropy of rocks and partly to rock testing methodology.

4-25. Poisson's Ratio

POISSON's ratio $\mu = 1/m$ or POISSON's number $m = 1/\mu$ is another important quantity in the theory of elasticity. POISSON's ratio μ is defined as the ratio of induced lateral strain ε_x to the longitudinal strain ε_z in the direction of applied axial load:

$$\mu = \frac{\varepsilon_x}{\varepsilon_z} = \frac{1}{m} \tag{4-57}$$

where $m = 1/\mu$ = POISSON's number.

For small deformations, a material whose POISSON's ratio is $\mu = 1/2$ is referred to as incompressible. POISSON's ratio for rocks varies according to the nature of deformation. Its value is usually relatively small. For sound and hard rocks within the elastic domain, POISSON's ratio is of the order of magnitude of about 0.15.

It increases near failure to about $\mu = 0.30$, and at constant deformation to approximately $\mu = 0.50$.

Elasticity constants of anisotropic crystals may be found in the *American Institute of Physics Handbook* (18).

Relative to rock materials in situ, it should also be said that the so-called elasticity constants such as the modulus of elasticity E and POISSON's ratio μ are not constants at all.

In reality, rather, these constants are variable quantities. This becomes clear if one remembers that rock is a nonhomogeneous, anisotropic material in different directions (rock defects). Therefore the modulus of elasticity, too, is different in different directions. These factors, as well as the effect of rock grain interlocks and wedging all affect the magnitude of POISSON's ratio, also.

To conclude the discussion on deformation of rocks, it should be said that because rock is not a perfectly linear-elastic material, certain allowances and intelligent choices must be made for deviations from perfect, ideal elasticity based on the engineer's personal experience, thus reconciling theory with reality and rendering a rational approach to design in rock.

4-26. Strength Properties of Rocks

For engineering design in rock, rock performance involves the determination of appropriate elastic constants and strength properties of the rock.

The ability of a material to resist externally applied forces is called its strength. In ordinary engineering practice, strength may be regarded as the force per unit area necessary to bring about rupture at given environmental conditions.

As applied to rock, the term "strength" is somewhat relative. It can be defined only when all the strength-governing factors such as rock environment; size of rock specimen; kind, intensity and duration of load; all-around lateral confining pressure; temperature; pore-water pressure; and failure criteria are known. The strength of small specimens of sound rock depends entirely on the strength of the rock matrix.

Needless to say, the strength properties of a rock are also governed by the qualitative and quantative mineral composition of the rock. Besides hardness of minerals and durability of rock, there are other factors, too, affecting the mechanical properties of rock. These are:

1. Type of rock;
2. Locality and environment of the rock;
3. Internal strength of individual mineral particles;
4. Strength of mutual bonding of the mineral grains of the rock;
5. Orientation of the mineral crystals and grains relative to loading and subsequent lateral deformation and/or sliding (this condition is especially pronounced with shale and slaty rocks easily cleavable and internally slidable minerals [mica]);
6. Various rock defects such as joints, cracks, haircracks, fissures, voids, gaps, and pores of all sorts;
7. Degree of rock saturation with water;
8. Coarseness of rock particle sizes and their strength;
9. Rock elasticity;
10. Plasticity;
11. Initial stress of rock in situ;
12. Testing methodology (16);
13. Rate of testing; and
14. Time.

One distinguishes between rock

1. Static strength, and
2. Dynamic strength.

The strength of a rock may be established either experimentally by means of laboratory testing of intact rock specimens, and/or by rock-testing in situ (4-9, 11-15, 28-32, 37, 41). Rock strength tests in situ are performed because a rock may contain various kinds of rock defects and planes of slippage (61).

Static Strength

Rock tests are made for the purpose of obtaining numerical values of their physical and strength properties to be used for design in rock.

The strength of a rock is determined in a variety of ways. For elastic materials, strength is described by means of the elastic constants E and μ, which are characteristic of each material.

Elastic constants are determined by static methods of test, or by means of dynamic methods; in the latter, velocities of propagated wave energy are measured.

Also, strength tests on rocks may be performed in the laboratory, and/or in situ (4-9, 11-15, 71-74, 76, 77, 84, 85, 87, 88, 91, 92).

Important static, laboratory-tested strength properties of rocks are:

1. Compressive strength:
   a. Uniaxial or unconfined compressive strength, and
   b. Triaxial compressive strength.
2. Direct shear strength (involves coefficient of friction f),
3. Tensile strength,
4. Bending or flexural strength, and
5. Punching strength.
6. Sometimes, also, simple torsion tests on rock are made.
7. Thermal stresses and strains,
8. Plastic properties, and
9. Creep.

4-27. Static Laboratory Compressive Strength

There exist unconfined and confined methods of compression tests.

The laterally *unconfined,* uniaxial compressive strength test is generally known to be one of the most rigorous strength tests of a material. It is also the commonest method for studying the mechanical properties of rocks.

The unconfined compression test is performed on cylindrical, or prismatic, or cube rock specimens (Fig. 4-19) (13) by compressing or loading them to failure. The commonly used height-to-diameter ratio of cylindrical rock specimens is $h/d = 2.5$ to 3.0. Upon failure, the rock specimen usually fractures by axial, brittle splitting, or fails in shear, depending upon the degree of the end contraints at the ends of the rock specimen offered by

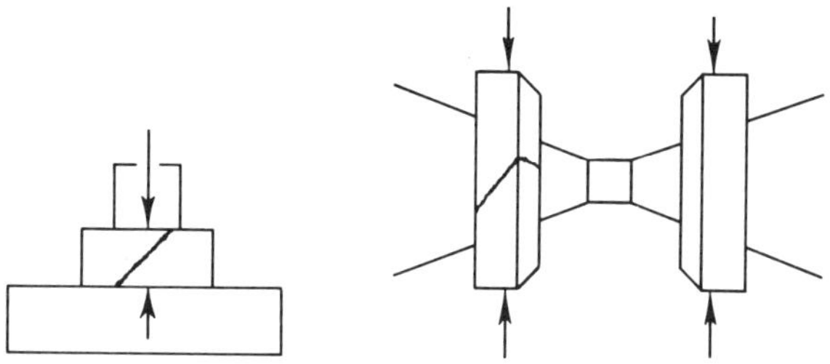

i) Loading of rock fragments

ii) Loading of rock pillars in an underground opening

a) Natural loading in situ

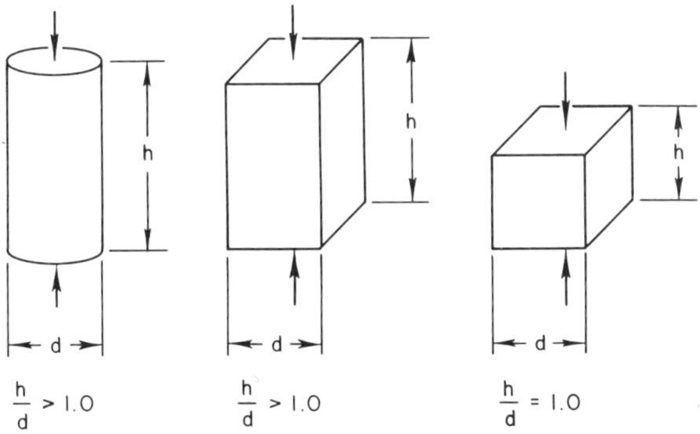

b) Unconfined compression testing of rock cylinders, prisms, and cubes in the laboratory

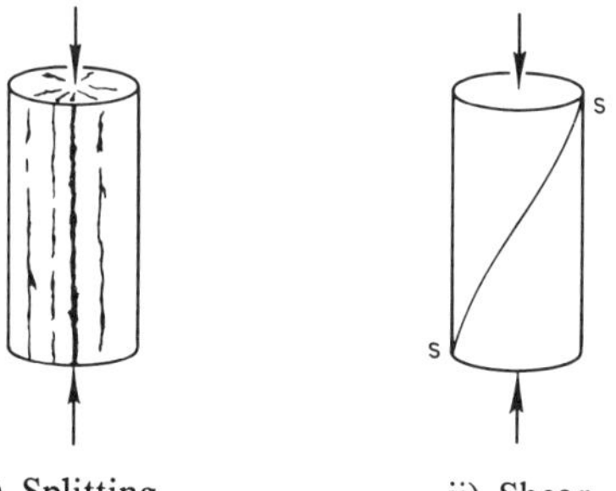

i) Splitting

ii) Shear

c) Modes of failure

Fig. 4-19
Rock test specimens for unconfined compression tests

the platens of the testing machine and the surface quality of the parallel ends of the rock specimen receiving the load (Fig. 4-19c).

Upon failure of the rock specimen at the ultimate compressive stress σ_1, when the shear strength in its potential shear plane *s-s* is exhausted, the sheared-off parts slide past each other across the inclined *s-s* plane; i.e., the rock fractures or fails in shear.

For any axial stress applied to the rock core specimen, the axial and lateral strains are measured either by strain gauges fitted on the test specimen, or by measuring displacements. These data are used for determining the POISSON's ratio of the tested rock specimen. Also, deformations are recorded for plotting of stress-strain diagrams to determine the modulus of elasticity E of the tested rock specimen.

The unconfined compression test results also render indirectly the unconfined shear strength $s = \tau$, and the test parameters φ and c (if any), commonly known as the angle of friction and cohesion respectively, as a function of rupture angle α (Fig. 4-20). Friction on the rupture surface supports a con-

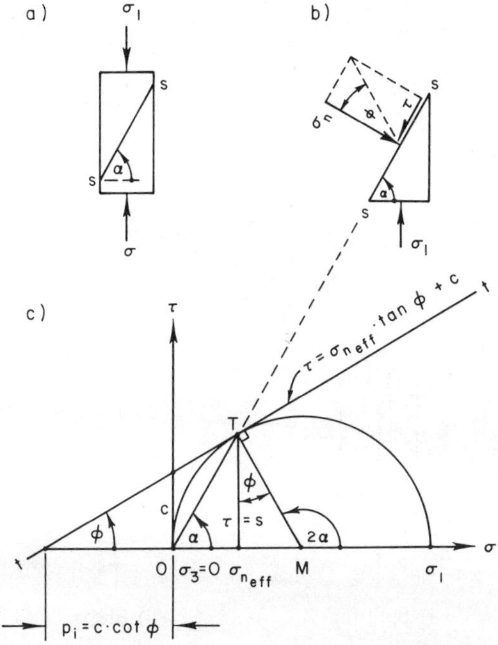

Fig. 4-20
Unconfined compression test

a) Failure of rock specimen
b) Free-body diagram of rock element
c) MOHR's stress diagram and COULOMB's shear strength envelope (tangent $t-t$)

siderable amount of axial stress. Cohesion may be regarded as the tangential strength (no-load strength).

The unconfined shear strength s of a rock is expressed by COULOMB's shear strength equation (Fig. 4-20):

$$s = \tau = \sigma_n \cdot \tan\varphi + c \tag{4-58}$$

When $c = 0$,

$$s = \tau = \sigma_n \cdot \tan\varphi \tag{4-59}$$

where

$$\sigma_n = \frac{\sigma_1}{2}(1 + \cos 2\cdot\alpha) \tag{4-60}$$

is the effective normal stress on the shear plane. From geometry of MOHR's stress diagram,

$$2\cdot\alpha = \frac{\pi}{2} + \varphi \tag{4-61}$$

and

$$\varphi = 2\cdot\alpha - \frac{\pi}{2} \tag{4-62}$$

The study of friction is of utmost importance in rock mechanics.

The test parameter φ, viz., $\tan\varphi$, the coefficient of friction, calculates from MOHR's stress diagram as

$$\tan\varphi = \frac{\tau}{\sigma_n + p_i} = \frac{\dfrac{\sigma_1}{2} - \sigma_n}{\tau} = f \tag{4-63}$$

where

$$p_i = c \cdot \cot\varphi \tag{4-64}$$

is the initial stress in the rock, and

$$f = \tan\varphi \tag{4-65}$$

is the coefficient of friction. From here,

$$\varphi = \arctan(f) \tag{4-66}$$

The cohesion c calculates from MOHR's stress diagram as

$$c = \tau - \sigma_n \cdot \tan\varphi \tag{4-67}$$

or

$$c = \tau \cdot \frac{p_i}{p_i + \sigma_n} \tag{4-68}$$

The *triaxial* compressive strength is tested by subjecting a rock specimen to a lateral, all-round confining pressure σ_3 and loading the specimen axially to failure (σ_1). This failure stress σ_1 is referred to as the triaxial compression strength of the rock specimen. The description of the general procedure for performing the triaxial compression test may be found in the book on ASTM standards (15). See also Ref. 79.

Also in the triaxial compression test, the rock specimen may fail either by axial, brittle splitting, or in shear by the characteristic cones formed on the shear surfaces.

Also, from the triaxial compression test data it is possible to evaluate the effective normal stress ($\sigma_{n\,eff}$) and the shear stress (τ) on the rupture plane, as well as the test parameters such as the angle of friction φ and cohesion c (if any) as a function of the angle of rupture α of the rock specimen. Thus, the triaxial compression test, too, gives indirectly the shear strength of the rock.

The shear strength s in this test is (Fig. 4-19):

$$s = \tau = \sigma_{n_{eff}} \cdot \tan \varphi + c \tag{4-69}$$

The φ and c values from a triaxial test are to be determined from the appropriate geometry of Mohr's stress diagram, as in Fig. 4-21.

Triaxial testing has the disadvantage that the geometry of the test specimen changes after any slip deformation, so that continued experiments of long duration are not very possible. However, for small displacements this test is considered to be adequate. It is the method most commonly used for determining the stress parameters φ and c.

Projecting forces acting on the free-body rock element for equilibrium obtain the normal (σ_n) and shear stress (τ) components in terms of the major (σ_1) and minor (σ_3) principal stresses, and as a function of α:

$$\sigma_n = \frac{\sigma_1 + \sigma_3}{2} + \frac{\sigma_1 - \sigma_3}{2} \cdot \cos 2\alpha \tag{4-70}$$

$$\tau = \frac{\sigma_1 - \sigma_3}{2} \cdot \sin 2\alpha \tag{4-71}$$

When $\sigma_3 = 0$ (as in the unconfined compression test), then

$$\sigma_n = \frac{\sigma_1}{2} + \frac{\sigma_1}{2} \cdot \cos 2\alpha = \frac{\sigma_1}{2} \cdot (1 + \cos 2\alpha) \tag{4-72}$$

Static compression tests on *cubes* render rock compressive strength only (12). Like the cylindrical specimens, so also the cubes fail either by brittle splitting, or in shear (Fig. 4-22).

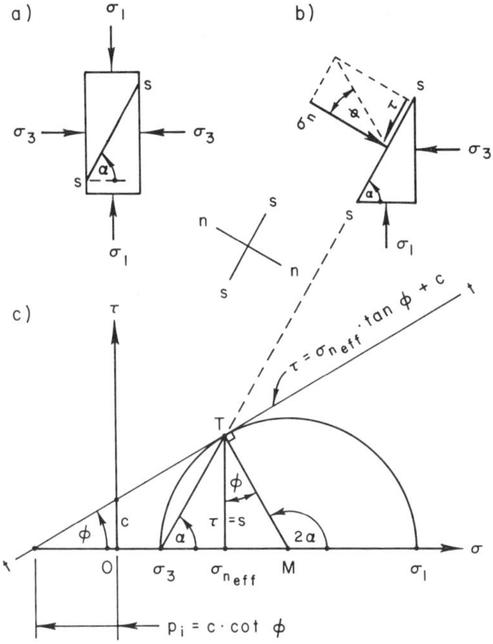

Fig. 4-21
Triaxial compression test
a) Failure of rock specimen in triaxial loading
b) Free-body diagram of rock element
c) MOHR's stress diagram and COULOMB's shear strength envelope ($t - t$)

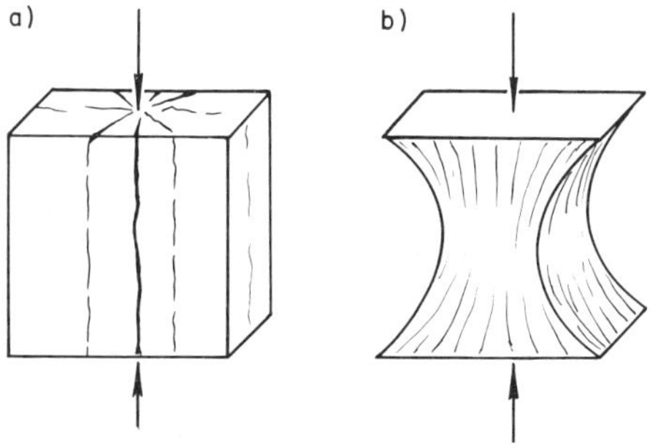

Fig. 4-22
Compression tests on rock cubes
a) Brittle splitting
b) Shear

Static tests which are performed during a short period of time and at relatively low loading rates (e.g., 10 psi/s = 68.94 (kN/m^2)/s to 100 psi/s = 689.4 (kN/m^2)/s) are referred to as *time-independent* as contrasted to *time-dependent* tests. Time-dependent tests pertain to creep tests under long sustained loading, for example.

The end faces of the test specimens must be strictly parallel to each other, and perpendicular to the generatrix of the rock cylinder, or prism, or cube.

The faces must be grounded to close tolerances. The following standards have been suggested by the International Society for Rock Mechanics' Committee on Laboratory Tests of Rocks, Document No. 1, of 1972 (43):

1. In strength tests of rock, the use of capping materials or end-surface treatments other than machining is not permitted. This is to be interpreted that the silicon grease method to eliminate the end friction effects of the rock specimens while testing is not permitted either.
2. The ends of the specimen shall be flat to 0.02 mm.
3. The ends of the rock specimen shall be perpendicular to the axis of the specimen within 0.001 radian (3.5 minutes).
4. The sides of the specimen shall be smooth and free of abrupt irregularities and straight to within 0.3 mm over the full length of the specimen.

Usually rock strength test results scatter within a wide range. This brings up the question as to how many specimens are necessary for such tests.

If identical compression tests are performed on *n* rock specimens, then the compressive strength σ is expressed as:

$$\sigma = \sigma_{\text{ave}} \pm \Delta\sigma \tag{4-73}$$

where

$$\sigma_{\text{ave}} = \frac{\Sigma\,(\sigma_1)}{n} \tag{4-74}$$

is the average strength of *n* specimens, σ_1 = ultimate compressive strength of an individual specimen, and

$$\Delta\sigma = \sqrt{\frac{(\sigma_1 - \sigma_{\text{ave}})^2}{n - 1}} \tag{4-75}$$

is the standard deviation computed by the method of least-squares.

Generally, the compressive strength of rock is a function of the following:

1. Specimen size,
2. Shape,
3. Surface quality of bearing platens and rock specimen end surfaces,
4. Confining pressure,
5. h/d-ratio, where h = height of core specimen and d = its diameter,
6. Rate of loading,
7. Porosity of rock, and
8. Moisture content.

As the confining pressure σ_3 is increased, the strength of the rock increases; also, the amount of the permanent set, or irreversible deformation, before fracture increases.

According to JAEGER (47), this behavior is characteristic of many rocks, even to confining pressures of the order of magnitude of 50 kpsi = 344.75 MN/m², that is, throughout the range of importance in practical rock mechanics.

If the rock specimens having the same diameter are of different length (height), the various ultimate compression-strength data from such specimens are adjusted, for the sake of uniformity of comparison, according to the *ASTM Standard Method of Test for Compressive Strength of Natural Stone* (12), to the strength of a rock specimen having a diameter-to-height ratio of $b/h = 1.00$, as follows:

$$\sigma_{cu} = \frac{\sigma_{ult}}{[0.778 + (0.222) \cdot (b/h)]} \tag{4-76}$$

where σ_{cu} = computed ultimate compressive strength of an equivalent cubical (or cylindrical) specimen whose ratio is $b/h = 1.00$, termed here the *adjusted compressive strength;* and

σ_{ult} = ultimate compressive strength of the test specimen whose height h is greater than its diameter $d = b$, i.e., whose b/h-ratio is less than 1; all in consistent units of measurement.

Here b is the width of the cube or prism. In the case of a cylindrical specimen, $b = d$, where d is the diameter of the rock specimen.

In Eq. (4-76), if $b/h = 1.00$, then $\sigma_{cu} = \sigma_{ult}$. Equation (4-76) also shows that the b/d-ratio, too, influences the compression strength test results. For example, for a constant-diameter rock specimen ($d = b =$ const), the strength decreases with increase in height of the specimen.

Generally, small b/h-ratio rock specimens fail because of elastic instability.

According to MOOS and QUERVAIN (65), water in rock pores reduces the magnitude of the internal friction of rocks.

Medium b/h-ratio ($\sim$ 0.3 to 0.4) rock specimens are elastically stable. In such specimens the stress distribution is fairly uniform.

Compressive strength of rock also decreases with increase in rock porosity, viz., volume of voids. This is so because at the location of voids there is a lack of bond contact between rock particles. However, at this point, it should be said that one cannot estimate the decrease in strength merely by the volume of the voids: the shape of the voids (narrow, rounded, or otherwise); the magnitude of the internal surface area of the grains because of the voids and gaps, each of which may be of different size; as well as the arrangement of the voids in the rock (interconnected or disconnected)—all these factors affect the strength of the rock.

Frequently one encounters coarse-textured granites which are very porous. Such rocks have considerably less strength than some fine-textured igneous rocks. In other words, the strength of a fine-grained rock is greater than that of a coarse-grained rock.

Basalt is one of the most competent of common rocks. It has a microfine texture, and consists of microcrystals held together by strong mechanical bonding.

Highly foliated schistose gneisses usually exhibit a wide range of strength.

Water in rock, too, affects the compressive strength and the elastic properties of rock. Generally, strength decreases with increase in moisture. Saturated rocks have less strength than dry ones. OBERT and DUVALL (73) are of the opinion that "a part of this decrease in strength with increase in moisture content may be due to an inability of the pore water to migrate freely (within the time limit of the test), and hence a pore (water) pressure may develop as the load on the specimen is increased."

The ratio of dry to wet strength is called the *softening factor*.

The author's experimental studies about New Jersey Triassic shale (55) revealed the great and relatively consistent decrease in unconfined compressive strength of the shale specimens upon soaking them for two weeks in water. The average value of soaked specimens was approximately one-third of the average compressive strength of oven-dry specimens. The softening factor S_f of dry to wet strength in these tests is:

$$S_f = \frac{\sigma_{\text{ult dry ave}}}{\sigma_{\text{ult wet ave}}} \approx 3.36 \tag{4-77}$$

Pore-water pressure in rock is a factor affecting rock strength. If the rock contains an interconnected system of water-filled pores, the fracture, viz., rock strength, is also governed by pore-water pressure, viz., effective stresses. The pore-water pressure reduces the principal stresses. In such a case, the effective principal stresses $\sigma_{1\,\mathrm{eff}}$ and $\sigma_{3\,\mathrm{eff}}$ are expressed as:

$$\sigma 1_{\mathrm{eff}} = \sigma 1_{\mathrm{total}} - u \tag{4-78}$$

$$\sigma 3_{\mathrm{eff}} = \sigma 3_{\mathrm{total}} - u \tag{4-79}$$

and the effective normal stress as:

$$\sigma_{\mathrm{n\,eff}} = \sigma_{\mathrm{n}} - u \tag{4-80}$$

where u = neutral stress known also by the term "pore-water pressure."

In terms of the effective stress, COULOMB's shear strength equation, Eq. (4-58), is written as:

$$s = (\sigma_{\mathrm{n}} - u) \cdot \tan \varphi + c \tag{4-81}$$

The effect of pore-water pressure on failure is shown in Fig. 4-23.

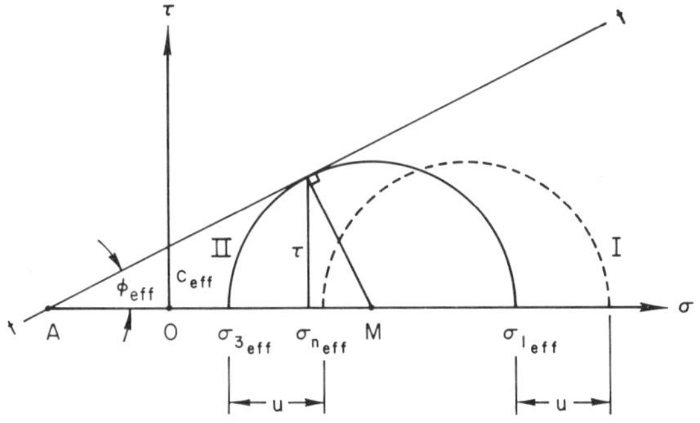

Fig. 4-23
Effect of pore-water pressure u on principal and normal stresses

Circle I is MOHR's stress diagram for actual, total stresses. Circle II is MOHR's stress diagram for effective stresses. Tangent t-t to the effective stress circle is COULOMB's strength envelope to the effective stress circle. As the pore-water pressure u is increased, stress circle I shifts to the left by an amount of u until it tangents the strength line to bring about failure. Notice that the stress difference $\sigma_1 - \sigma_3$, and the slope of the tangent, or envelope, remain unchanged.

Some compressive-strength values of rocks are compiled in Table 4-10 from various sources (32, 86).

TABLE 4-10

Compressive Strength σ_c, Tensile Strength σ_t, and Shear Strength $\tau = s$ of Various Kinds of Rocks

Rock	Dry, Unconfined Compressive Strength, σ_c		Tensile Strength, σ_t		Shear Strength, $(\tau = s)$		References
	kg_f/cm^2	MN/m^2	kg_f/cm^2	MN/m^2	kg_f/cm^2	MN/m^2	
(1)	(2)	(3)	(4)	(5)	(6)	(7)	(8)
Igneous rocks							
Basalt	800–4200	78–412	60–120	5.9–11.8	50–130	4.9–12.7	86
	1500–3000	147–294	100–300	9.8–29.4	200–600	19.6–49.0	32
Diabase	1200–2500	118–245	60–130	5.9–12.7	60–100	5.9– 9.8	86
Gabbro	1500–2000	147–196	50– 80	4.9– 7.8	40– 85	3.9– 8.3	86
	1800–3000	177–294	150–300	14.7–29.4	–	–	32
Granite	1200–2800	118–275	40– 80	3.9– 7.8	50–100	4.9– 9.8	86
	1000–2500	98–245	70–250	6.9–24.5	140–500	13.7–49.0	32
Sedimentary rocks							
Dolomite	150–1200	14.7–118	25– 60	2.5– 5.9	25– 70	2.5– 6.9	86
	800–2500	78–245	150–250	14.7–24.5	–	–	32
Limestone	40–2000	3.9–196	10– 70	1.0– 6.9	15– 70	1.5– 6.9	86
	300–2500	29.4–245	50–250	4.9–24.5	100–500	9.8–49.0	32
Sandstone	600–1000	49.0– 98	20	19.6	30	2.9	86
	200–1700	19.6–167	40–250	3.9–24.5	–	–	82
Shale (clay)	220–1635	21.6–160	–	–	30–110	2.9–10.8	86
	100–1000	9.8– 98	20–100	2.0– 9.8	30–300	2.9–29.4	
Sandstone	500	49	–	–	–	–	32
Methamorphic rocks							
Gneiss	800–2500	78–245	40– 70	3.9– 6.9	30– 70		86
	800–2000	78–196	80–200	7.8–19.6	–	–	32
Marble	500–1800	49–177	50– 80	4.9– 7.8	35– 80		86
	1000–2000	98–196	70–200	6.9–19.6	150–300		32
Quartzite	870–3600	85–353	30– 50	2.9– 4.9	–	–	86
	1500–3000	147–294	50–200	4.9–19.6	200–600	19.6–58.8	32
Slate	250– 800	24.5– 78	–	–	–	–	86
	1000–2000	98–196	70–200	6.9–19.6	–	–	86

86. Széchy, K., 1966, *The Art of Tunneling*. Budapest: Akadémiai Kiadó Budapest, p. 76.
32. Farmer, J. W., 1968, *Engineering Properties of Rocks*. London: E. and F. N. Spon, Ltd., p. 57.

4-28. Tensile Strength of Rock

According to the ASTM D 653-67 Standard Definitions of Terms and Symbols Relating to Soil and Rock Mechanics (10), the tensile strength of a material is defined as "the maximum tensile stress which a material is capable of developing."

Practically, the tensile strength is considered to be the maximum stress developed by a given specimen of a material in a tension test performed to rupture under specified conditions (14).

In rock mechanics, knowledge about tensile strength of rocks is important in analyzing rock strength and stability of roofs and domes of underground openings in the tensile zone of the rock (Sections 7-4 through 7-6, and 11-5), in mining for minerals, in preparing rock drilling and blasting programs, and possibly in other endeavors in rock engineering.

Tensile stresses may be induced on the underside of a rock slab or beam subjected to bending, and on the leeward side of a buckling rock pillar (Figs. 4-24a and b). A piece of rock may also fail in tension, as shown in Fig. 4-24c.

It is generally known that, because of the discontinuities a rock may contain, rock in its natural state is relatively weak in tension. The tensile strength σ_t of a rock is much less than its compressive strength σ_c. The tensile strength of rock is only about 10% that of its compressive strength, i.e.,

$$\sigma_t \approx (0.10) \cdot \sigma_c \tag{4-82}$$

As compared with compressive strength, rock tensile strength has not been studied extensively enough. One possible explanation for this may be the difficulty in preparing a tensile test specimen of rock; fixing it properly in the chucks of the testing machine; and centering exactly the axial, tensile load. Thus, tensile strength tests on rocks are difficult to perform.

To determine the tensile strength σ_t, the rock specimen is subjected to maximum axial tensile force P attained during the test to failure. During the loading process, the strains are measured. Then the corresponding stresses and strains are computed and a stress-strain diagram is plotted. The average tensile strength σ_t is computed as:

$$\sigma_t = \frac{P}{A_o} \tag{4-83}$$

where A_o = initial, unnecked, cross-sectional area of the rock test specimen.

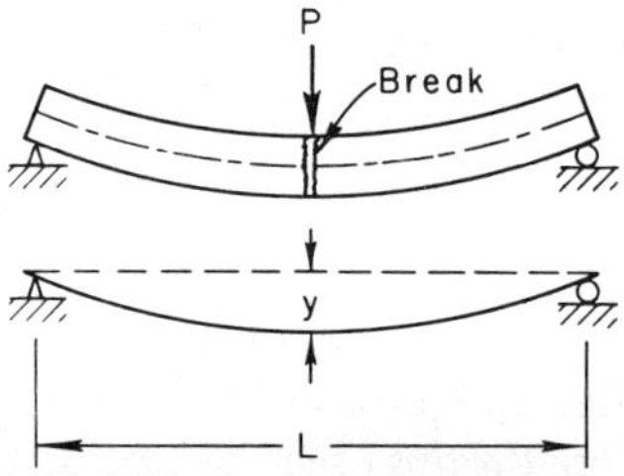

a) Bending of beam

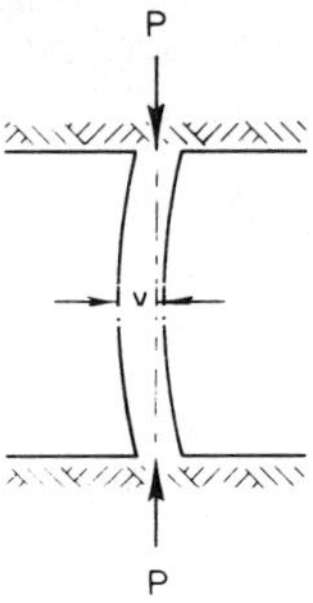

b) Buckling of column

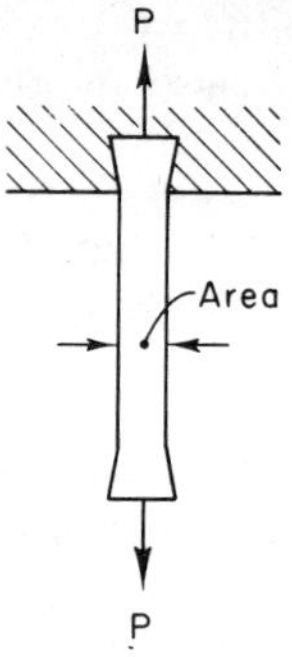

c) Loading in tension

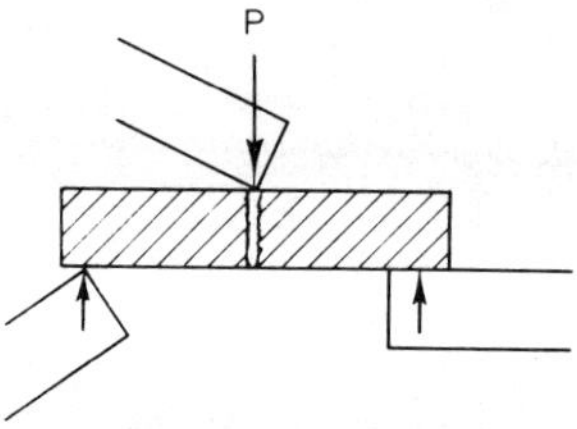

d) Failure of rock in bending

Fig. 4-24
Some modes of deformation of materials in tension

In this test, failure occurs on the weakest cross-sectional plane (such as a bedding plane), or a parting, or a gouged or weakly cemented joint of the rock specimen.

So far, systematic studies on rock *flexural (bending) strength* have been performed very seldom. Flexural strength tests are performed for purposes of determining YOUNG's modulus of elasticity E, and tensile strength of the material.

Flexural strength tests on rock are performed on a test beam made of rock. The simply supported beam at its ends is loaded with a concentrated load P at its midpoint to failure. In such a loading, the underside fibers of the beam are under tension.

From strength of materials, the maximum bending (flexural) tensile stress σ_t at the extreme fiber on the underside of a rock beam is considered to be the tensile strength σ_t of the rock:

$$\sigma_t = -\frac{M}{I/c} = -\frac{M}{Z} \quad (4\text{-}84)$$

$$= -\frac{P \cdot \left(\frac{L}{2}\right)}{I/c} = -\frac{P \cdot c \cdot L}{2 \cdot I} \quad (4\text{-}85)$$

where M = bending moment,
I = central moment of inertia of cross-section of the beam,
c = distance from neutral axis to the farthermost fiber,
$Z = I/c$ = section modulus of the beam, and
$L/2$ = one-half of the span of the beam.

The quantity σ_t is also known as the modulus of rupture.

It is generally known that the tensile strength of rocks obtained from flexure tests on rock beams is considerably larger than its uniaxial tensile strength.

The average modulus of elasticity E_{ave} for a central deflection y of the beam is:

$$E_{ave} = \frac{P \cdot L^3}{6 \cdot I \cdot y} \quad (4\text{-}86)$$

For a rectangular cross-sectional beam, $I = b \cdot h^3/12$. For a circular cross-sectional beam, $I = \pi R^4/4$.

In order to take into account the difference in YOUNG's modulus of elasticity in tension and compression, Duckworth (26) put forward for rectangular, ceramic beams the following equation for determining tensile strength σ_t:

$$\sigma_t = -\frac{3M(\varepsilon_1 + \varepsilon_2)}{b \cdot t^2 \cdot \varepsilon_1} \tag{4-87}$$

Here, M = bending moment,
ε_1 = tensile bending strain of the outer fiber,
ε_2 = compression bending strain of the inner fiber,
b = width of the gauge section, and
t = thickness of the gauge section.

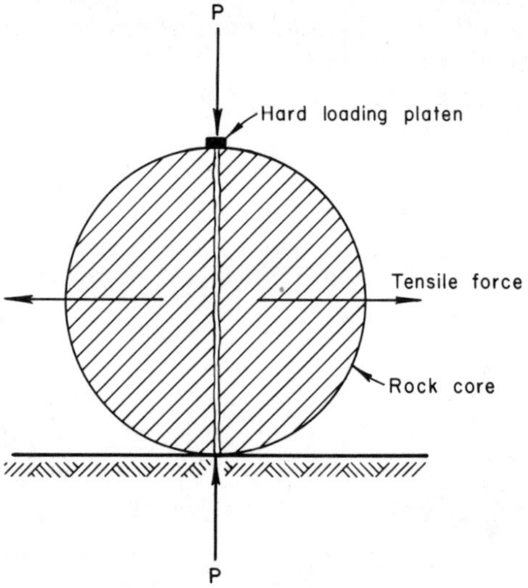

Fig. 4-25
Indirect (Brazilian) tensile strength test

Another, indirect method of testing rock tensile strength is the so-called *Brazilian test* (30, 32, 46-48, 91). In this test, a cylindrical rock specimen, lying on its side, is loaded diametrically with a compression load P, as sketched in Fig. 4-25, to bring about a uniformly distributed tensile stress over the vertical, central, diametrical plane. Ideally, failure in tension should take place as a clean break. The tensile strength σ_t of the rock in this test is:

$$\sigma_t = \frac{2 \cdot P}{\pi \cdot d \cdot h} \tag{4-88}$$

where d = diameter of specimen, and
h = height (or length) of the specimen.

For one's orientation, some tensile strength and bending strength values for various kinds of rocks are compiled from various sources in Table 4-10.

4-29. Shear Strength of Rock

There are many design problems in rock mechanics where knowledge about the shear strength of a rock mass is needed. Among them are:

1. Stability problems of underground openings,
2. Assessment of the degree of stability of rock slopes, and
3. Evaluation of stability of a structure against sliding on its base laid on rock.

Some modes of shear failure in rock are sketched in Fig. 4-26.

The shear strength parameters to use in stability calculations of a rock mass and/or structure, in or on rock, point to the kind of shear strength test to be performed (55, 74, 76, 77, 85).

The shear strength of rock is its maximum resistance s to deformation by continuous shear displacement upon the action of a shear (tangential) stress τ.

Upon exceeding the shear resistance, or strength s, failure of rock in shear takes place.

The shear strength s of a rock is the sum of:

1. The surface frictional resistance to translation in the sliding surface,
2. The interlocking effect between the individual rock grains, and
3. Cohesion, if any, in the sliding surface of the rock.
   (Cohesion, it may be said, is the no-load shear strength of the rock.)

When motion, viz., displacement in shear, is just impending, there exists a state of static equilibrium, i.e., the shearing stress τ is equal in magnitude and oppositely directed to the shear strength s:

$$\tau = s \tag{4-89}$$

One should be cognizant that the applied stress and the strength stress in these tests are average, uniform stresses, and that they are *statistical averages* that do not represent the actual stresses on the microscopic parts of the rock. The stress is, as it were, "ragged" within microscopic irregularity within each kind of rock.

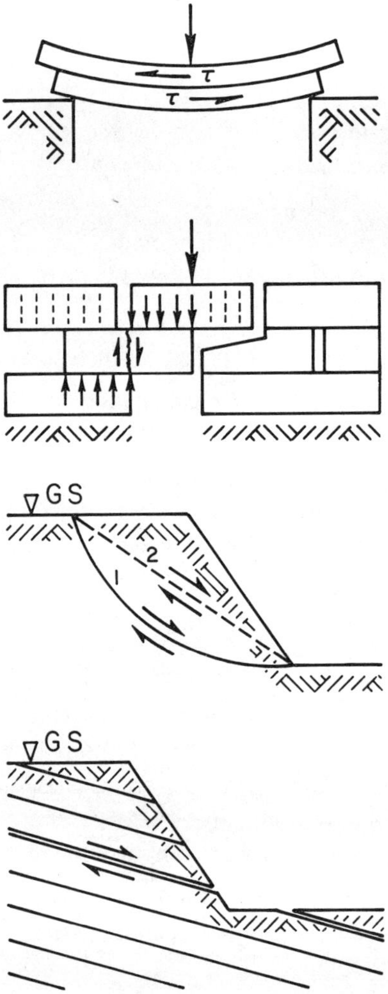

a) Shear between bending rock slabs: longitudinal shear displacement

b) Shearing a rock slab

c) Shearing of a homogeneous soil or rock slope:
1. Curvilinear shear surface
2. Plane shear surface

d) Shear in a pre-existing plane of weakness in rock

Fig. 4-26
Some modes of shear failure in rock

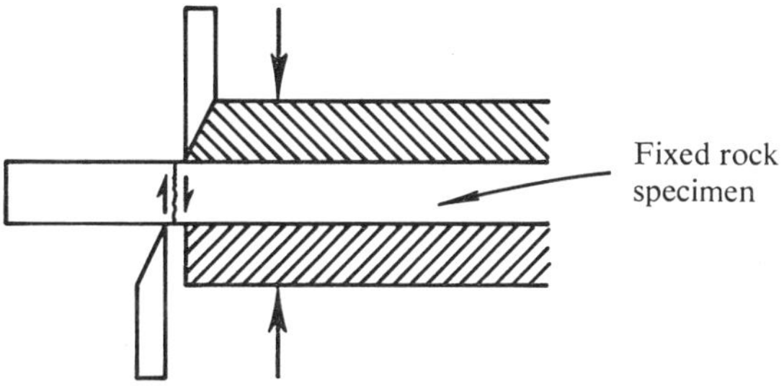

a) Single shear

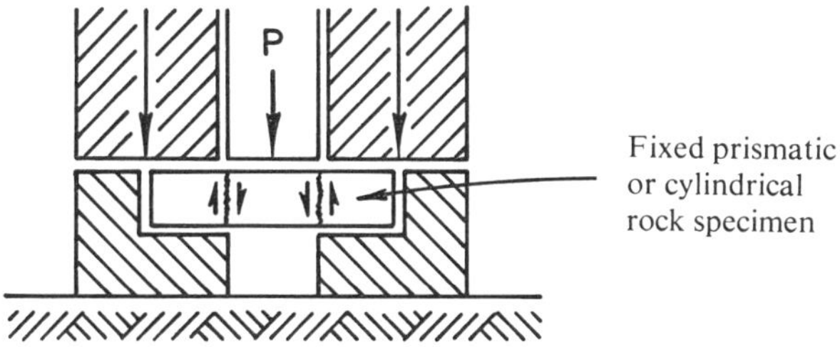

b) Double shear

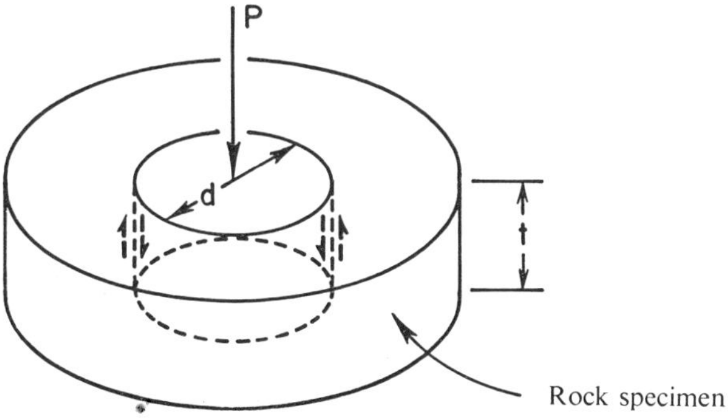

c) Punch shear

Fig. 4-27
Some principles of direct shear

Methods of Shear Testing of Rocks in the Laboratory

All shear tests on rocks may be grouped in the following groups:

A. Direct shear strength tests with normal stress on the shear plane absent,

B. Direct shear strength tests with normal stress on the shear plane present, and

C. Torsion tests.

A. Direct Shear Strength Tests with Normal Stress on the Shear Plane Absent

Because in this kind of test the normal stress on the shear plane is zero ($\sigma_n = 0$), the test is also termed the "unconfined" direct shear test. To this kind of test, there belong the following three kinds of direct shear tests:

1. Direct shear test in single shear,
2. Direct shear test in double shear, and
3. Punch shear.

The principles of these direct shear tests are illustrated in Fig. 4-27.

1. Single-Shear Test

The direct shear strength in single shear may be determined by placing a rock specimen between knife edges and exerting through them colinear pressure P on the specimen (Fig. 4-27a). The specimen fails along a predetermined, forced shear plane. The shear strength s of the rock is

$$s = \frac{P}{A} \tag{4-90}$$

where A = cross-sectional area of the shear plane of the tested rock specimen.

2. Double-Shear Test

This kind of uniaxial shear test is sketched in Fig. 4-27b. The prismatic or cylindrical rock specimen is fixed firmly into the specimen-holding jig. A shear force P is applied to the specimen through a die, bringing about shear failure in double shear. The shear strength s calculates as:

$$s = \frac{P}{2 \cdot A} \tag{4-91}$$

where A = cross-sectional area of the rock specimen.

3. Punch Shear

For the punch shear test (Figs. 4-27 and 4-28), flat rock disks are fitted into the testing jig (Fig. 4-28). The test is commenced by placing a rock

Fig. 4-28
Device for testing rock disks in punch shear

specimen at the bottom of the casing of a piston-shaped cylindrical jig having a puncher head which fits into a circular shear hole of another hollow cylinder. Then the puncher is placed in position and the entire device is placed in the compression-testing machine and loaded. The punching shear strength s of the rock specimen is:

$$s = \frac{P}{A} = \frac{P}{\pi \cdot d \cdot t} \tag{4-92}$$

where P = ultimate punching load,
A = circumferential shear area of punched-out circle of the thickness t of the disks, and
d = diameter of the puncher.

a) Shale disks before test

b) Shale disks after test

Fig. 4-29
Shale disks for punch shear tests
(Author's study, Ref. 55)

Some Triassic-shale disks after failure in punch shear resulting from the author's own experimental studies are shown in Fig. 4-29 (55).

Shale disk specimens tested out with the following average punching shear strength values:

$$\text{Dry:} \quad s_{\text{ave}} = 337 \text{ kg/cm}^2$$
$$\text{Wet:} \quad s_{\text{ave}} = 114 \text{ kg/cm}^2$$

The dry/wet factor was:

$$S_t = \frac{337}{114} = 2.96$$

Hence the dry specimens were approximately trice as strong in punch shear as the wet ones. After the punching tests, the deformed shale disks showed radial deformation patterns.

The advantage of the disk-punching method of shear strength is that it is easy to perform. This kind of test may be considered as a realistic representation of punching through a hard shale underlaid by a softer shale stratum.

A disadvantage of this kind of test is that the shear failure process cannot be observed. Also, in the punching shear test, the real stress distribution along the shear surface is very complex, and the shale test results scatter widely, deviating from the average from $\pm$ 28 to $\pm$ 37 %. Hence only the average punching shear strength data for the shale could be obtained.

B. Direct Shear Strength Tests with Normal Stress on the Shear Plane Present

Among these direct shear strength tests, one distinguishes between:

1. The so-called "box" shear strength tests on rock, and
2. Shear strength tests made on rock cubes.

The shear deformation in these tests take place along a predetermined, forced shear plane.

1. Direct "Box" Shear Strength Test on Rock

The so-called "box" shear strength test involves the fitting, snugly, of a rock specimen into a direct shear-testing box (Figs. 4-30a and b). The box consists of two frames, an upper and a lower. The rock specimen is being loaded with a vertical, normal load N, viz., normal stress $\sigma_n = N/A$, where A = area of shear plane. During the test, the normal load N is kept constant; a horizontal shear (tangential) force T, viz., shear stress $\tau = T/A$, is applied to the upper frame of the shear box; the shear force increments of T are recorded; and the corresponding horizontal displacements are measured. At failure, the shearing stress τ is equal to the shear strength s of the rock.

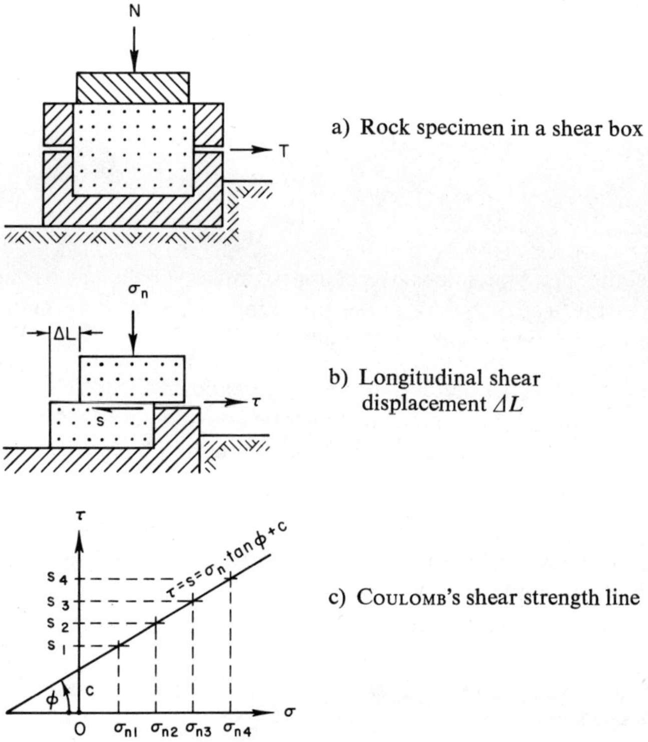

Fig. 4-30
Direct "box" shear strength test

The plot of $T = f(N)$, or $\tau = f(\sigma_n)$, results in COULOMB's straight-line shear strength relationship (Fig. 4-30c).

The straight-line relationship between the horizontal shear force T and vertical normal load N is given as:

$$T = N \cdot \tan \varphi + c \cdot A \qquad (4\text{-}93)$$

or

$$\tau = s = \sigma_n \cdot \tan \varphi + c \qquad (4\text{-}94)$$

where the shear strength test parameters are:

$\tan \varphi =$ coefficient of friction of the shear surface,
$\varphi =$ angle of friction,
$c =$ cohesion (if any), and
$A =$ area of shear.

Cohesion, it may be said, is the no-load shear strength.

If $c = 0$, then

$$\tau = s = \sigma_n \cdot \tan \varphi \tag{4-95}$$

The $\tan \varphi = f$ value depends upon the quality of the sliding, viz., contact surface between the mutually displacing upper and lower parts of the rock specimen deforming in shear, i.e., whether smooth or rough, level or inclined, dry or wet, quickly or slowly sheared.

If the rock specimen is tested along a forced inclined plane (Fig. 4-31a), the shear strength of the specimen is written as:

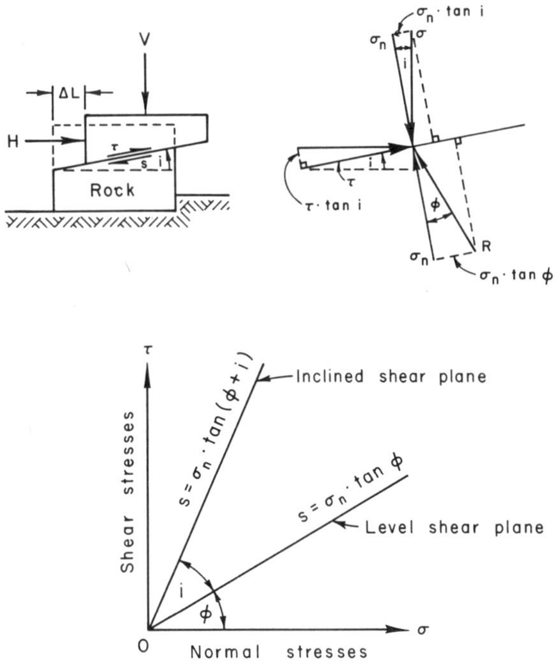

Fig. 4-31
Direct shear test along a forced inclined plane

$$s = \sigma_n \cdot \tan(\varphi + i) \tag{4-96}$$

where i = angle of slope with the horizontal of the forced inclined shear plane.

For such a case, the corresponding plot of $\tau = f(\sigma_n)$ is shown in Fig. 4-31b.

If the direct shear test is performed in water under submerged conditions, then the total hydrostatic uplift or neutral force U reduces the total normal force. Therefore the frictional resistance to sliding is also reduced. Hence, the shear force $\overline{T}$ for impending sliding is:

$$\overline{T} = (N - U) \cdot \tan \varphi \tag{4-97}$$

Expressed in terms of effective stresses,

$$\bar{\tau} = (\sigma_n - u) \cdot \tan \varphi \tag{4-97a}$$

$$= \sigma_{n_{eff}} \cdot \tan \varphi \tag{4-97b}$$

$$= \bar{\sigma}_n \cdot \tan \varphi \tag{4-97c}$$

where $\bar{\tau}$ = shear stress on effective stress basis,
σ_n = normal stress, and
$\bar{\sigma}_n$ = $\sigma_n - u = \sigma_{n_{eff}}$ = effective normal stress.

The magnitude of the total normal load N and the total neutral pressure U acting on the failure surface must be determined from rock exploration work in the field.

When working in the laboratory with a controlled-strain, direct shear testing device, or when working in situ, the test data for each single run of the test result in a shear stress-displacement diagram, as shown in Fig. 4-32a.

The angle φ_r of residual shearing resistance (Fig. 4-32b) is frequently used in stability calculations of jointed rock masses against their sliding along planes of weakness (74).

2. *Direct Shear Strength Tests of Rock Cubes*

This test is performed on prepared rock cubes. The loading of the cubes in this direct shear test is accomplished by means of a universal compression testing machine and a special device or jig (rock cube specimen holder, Fig. 4-33a) (55). The rock cube is placed in the holder diagonally at an angle of $\alpha = 45°$ with respect to the horizontal. Other jigs are designed for other angles with the horizontal (Fig. 4-33b). The bottom parts of these jigs are mounted with two loops of spherical bearings as a movable displacement base.

In the 45-degree jig, the loading device transfers the vertical load P to the rock specimen, forcing the rock cube to shear and fail in shear along a predetermined, forced-shear plane *s-s* (Fig. 4-34).

The normal stress σ_n is here calculated as:

$$\sigma_n = P \cdot \sin 45°/A = N/A \qquad (4\text{-}98)$$

and the shear stress τ as:

$$\tau = P \cdot \cos 45°/A = T/A \qquad (4\text{-}99)$$

(see Fig. 4-34), where N and T are forces on the shear plane and A ist the area of the shear plane.

This kind of direct shear test is designed to bring out the shear strength of the rock cubes parallel and perpendicular to the laminarity of the rock. Thus this test is a forced direct shear test.

Figure 4-35 shows some shale specimens after failure in the 45-degree direct shear test.

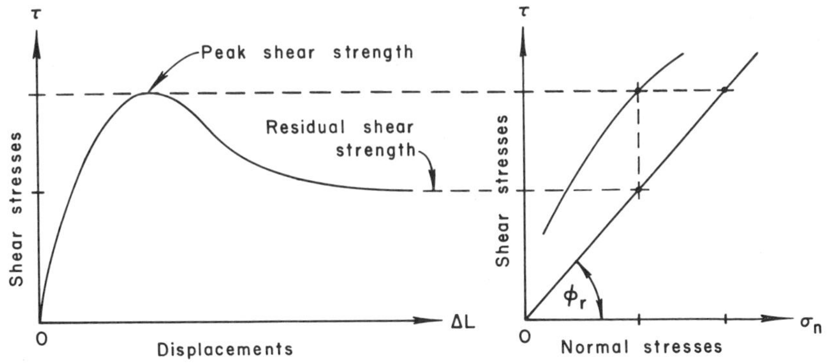

a) Shear strength as a function of displacement ΔL at constant normal stress σ_n

b) Angle φ_r of residual shear resistance

Fig. 4-32
Shear stress-displacement diagram
and angle of residual shear resistance φ_r

a) Rock cube holder for a
45-degree forced-failure
plane with the horizontal

b) Rock cube holders for various
angles of forced-failure plane
with the horizontal

Fig. 4-33
Devices for testing rock cubes in direct shear
(Author's study, Ref. 55)

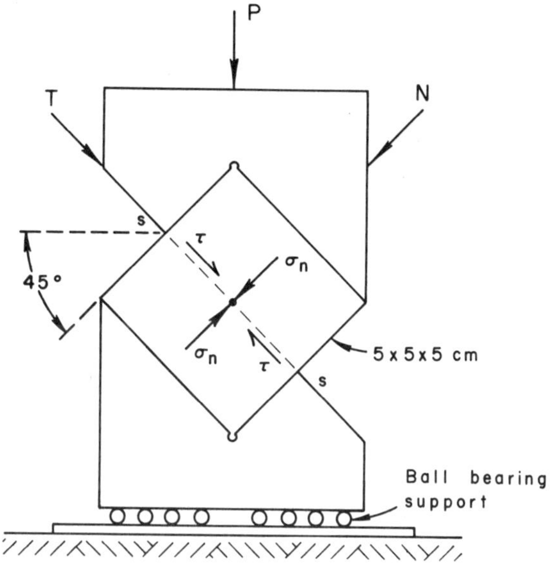

Fig. 4-34
Schematic rendering of a rock cube specimen holder
(Author's study, Ref. 55)

With jig inside inclinations other than 45 degrees, the rock can be forced to shear at different angles with the horizontal.

C. Torsion Tests

In the torsion tests, a solid-shaft cylindrical rock specimen, provided with square, prismatic ends for easy fixing into the chucks of the torsion-testing machine, is subjected to torsion or twisting. Upon torsion, the shear stress in the circular-shaft cross-section increases. It increases linearly from zero at the center of the circular cross-sectional area of the rock specimen to a maximum value τ_{max} at the outermost fiber of the shaft, where $r = r$ (Fig. 4-36):

$$\tau_{max} = \frac{(16) \cdot T}{\pi \cdot d^3} \tag{4-100}$$

where $T = P \cdot (2 \cdot r) =$ torsion moment (applied force couple),
$P =$ force,
$d = 2 \cdot r =$ diameter of the solid-shaft rock specimen.

Fig. 4-35
Shale cube specimens after failure in direct shear test
(Author's study, Ref. 55)
Top: wet specimens
Bottom: dry specimens

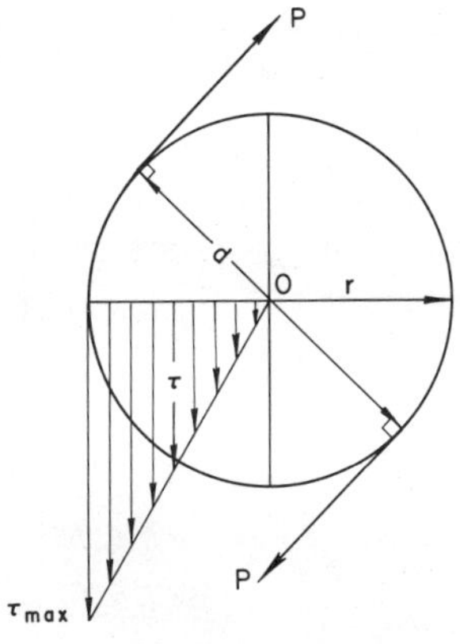

Fig. 4-36
Distribution of shear stresses in a cross-section of a circular, solid-shaft test specimen

For a rectangular shaft, a satisfactory formula for computing the maximum shear stress is (81):

$$\tau_{max} = \frac{T}{a \cdot b^2} \cdot \left(3 + 1.8 \cdot \frac{b}{a}\right) \tag{4-101}$$

where a = long side of the rectangular section, and
b = its short side.

For one's orientation, some shear strength values of various rocks are compiled from various sources in Table 4-10.

Friction

Resistance to sliding, viz., shear parallel to a geological discontinuity, is almost entirely frictional in character. Therefore, in rock mechanics and foundation engineering, the study of friction is of paramount importance (74, 91).

Friction occurs:

1. Between opposing surfaces of minute cracks in rock;
2. Between individual rock and soil particles; and
3. On joint or fault surfaces, small as well as on very large ones.

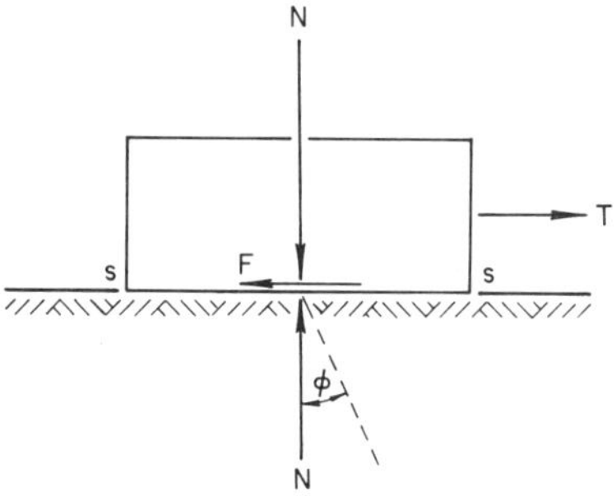

Fig. 4-37
Block at static equilibrium

So far, few studies have been devoted to friction on minerals and rocks.

From statics, at equilibrium (Fig. 4-37), the force of shearing resistance F is expressed as:

$$F = N \cdot f = T \tag{4-102}$$

where N = normal force of the block (including its own weight) on the contact surface s-s, whose area is A;

$f = \tan \varphi$ = coefficient of static friction; here f depends upon the quality of the contact surfaces, such as new or old surfaces, smooth or rough, for example;

T = externally applied tangential or shear force parallel to the surface of contact s-s necessary to initiate a sliding movement;

$F = N \cdot f$ = resistance to shear force T.

The normal force N presses the surfaces in contact together.

Upon dividing of the F-equation by the surface area A of contact, obtain the shear stress τ to bring about sliding:

$$\tau = f \cdot \sigma_n \tag{4-103}$$

Herein σ_n = normal stress on the contact surface s-s.

When sliding is in progress at a constant speed, the resisting force to sliding is also proportional to the load, but is independent of the area of the contact surface. In the case of sliding, the constant of proportionality is f'; this coefficient f' is called the coefficient of dynamic friction. The magnitude of the coefficient of dynamic friction is less than that of static friction, i. e.,

$$f' < f \tag{4-104}$$

The coefficient f' varies with the velocity of the sliding motion.

The values of coefficients of friction are determined in the laboratory by means of compression tests, as well as in situ.

If in the shearing plane s-s there is neutral pressure u present, then the principal and normal stresses are to be decreased by u to perform calculations with effective stresses.

In Table 4-11 are compiled coefficients of friction f as found in the various sources of literature by various researchers using different methodology and different types of equipment (18, 32, 47, 83).

4-30. Dynamic Properties of Rocks

In the laboratory, dynamic properties of rocks are determined on cubes by means of an impact device. One such dynamic property of rock is the *specific impact work* measured per 1 cm^3 of rock.

Frequently, the so-called *toughness index* I_T of rock is determined:

TABLE 4-11

Angles of Friction φ and Coefficients of Friction f

Rock	Angle of Friction φ, degrees	Coefficient of Static Friction $f = \tan \varphi$	References
Igneous rocks			18, 32, 47, 83
Basalt	48–50	1.11–1.19	
Diabase	50–55	1.19–1.43	
Gabbro	10–31	0.18–0.60	
	11.3–31	0.20–0.60	
Granite	56–58	1.48–1.60	
	45–60	1.00–1.73	
Sedimentary rocks			
Dolomite	22	0.40	
Limestone	35–50	0.70–1.20	
Sandstone	27–34.2	0.51–0.68	
	26.6–35	0.50–0.70	
Shale	15–30	0.27–0.58	
Metamorphic rocks			
Gneiss	31–35	0.60–0.70	
Marble	32–37	0.62–0.75	
	35–50.2	0.70–1.20	
Quartzite	25.6–60	0.48–1.73	
	50–60	1.20–1.73	
Schist	62.25	1.90	73

$$I_T = \frac{\text{specific impact work}}{\text{compressive strength}} \left[\frac{(N \cdot \text{cm}) \cdot \text{cm}^{-3})}{N \cdot \text{cm}^{-2}}\right] \qquad (4\text{-}105)$$

In these tests, a falling weight of 50 $kg_f \approx 490$ N is usually used. At first, 1 cm is chosen as the height of fall. In consecutive tests, the height of fall is increased by 1 cm.

However, the results obtained from these tests scatter very much: deviations of 50 to 100 % from the average are almost a normal occurrence. Therefore, perhaps only series of a large number of tests may somehow render statistical, comparable results. A disadvantage of these multinumber tests is their high cost. For this reason, such tests are very seldom performed.

Recently, studies on elastic properties of rocks (obsidian, in this case) have been attempted by the method of ultrasonic interferometry (62).

4-31. Summary on Laboratory Testing of Rocks

Rock tests performed in the laboratory are an important part of the engineering discipline of rock mechanics. The information needed for design in rock is obtained by direct testing, either on laboratory specimens or in the field. Much information can be obtained from laboratory tests of rock specimens with known orientation.

Laboratory testing of rocks, model studies, and indoor laboratory experimental work have the advantage over field work in that:

1. Loads, pressure, time, temperature, thermal gradients, and water can be controlled easily, whereas these and other factors for most rocks in their natural state in situ can be evaluated only crudely.
2. Laboratory testing equipment is convenient to use.
3. Laboratory tests are easy to perform; therefore, great precision is possible.
4. A laboratory-size rock specimen can be imposed easily upon an unconfined or triaxial load, and a sonic or shock wave propagated through it.
5. Laboratory tests enable one to obtain good and relatively inexpensive compressive strength data and shear strength parameters of intact rock.
6. Laboratory tests complement the test data as obtained from in-situ tests of rock.

Laboratory testing of rocks has also some disadvantages over in-situ testing, namely:

1. Disturbance is imparted on rock samples upon their sampling while separating them from the rock massif in situ.
2. There is possible disturbance, to some degree, of the structure of rock samples by shocks during their shipment from the sampling site to the laboratory.
3. To some extent, there is disturbance of rock properties during preparation of test specimens (sawing, coring, lapping, for example).
4. The strength properties of an intact laboratory rock specimen may differ greatly from the strength of the in-situ rock from which the specimen was sampled. It should be realized that the strength of a rock mass in situ, besides its own inherent intact strength, depends also to a high degree upon its various geological defects (viz., directional anisotropy). The effect of the defects of the prototype rock cannot be comprehended by laboratory tests on intact rock specimens free of fissures. Therefore, laboratory rock specimen test results do not represent the in-situ properties of the overall rock mass.

5. Laboratory test results do not shed any light about the internal stress conditions, primary or secondary, in the in-situ rock.

However, correct evaluation of rock strength properties obtained from appropriate laboratory and in-situ tests, supplemented with a high degree of experience and sound engineering judgment, contributes to safe and economic design of structures in and on rock.

4-32. Properties of In-Situ Rocks

A perfectly elastic solid is truly an ideal material. However, in contradistinction, rocks in situ contain various structural imperfections (rock defects and various planes of weakness). Obviously, rock defects prevent rocks from behaving in a perfectly elastic manner. For these reasons, it is said that rocks are inelastic materials behaving inelastically under loading and unloading conditions.

Because rock in situ is fissured through and through, it possesses entirely different strength properties than those obtained from intact rock core testing in the laboratory. Laboratory test samples are almost invariably taken from rock between major discontinuities and are, therefore, not representative of the rock mass.

Elastic properties determined from testing rock core specimens are reliable only for low stress levels. For high stresses, the jointing and seam structure and the anisotropy of the rock must be considered. Also, the plastic creep from rock crack and crystal contact adjustment with time caused by various kinds of stresses in rock should be taken into account.

Nearly every problem in rock engineering involves either the compressibility of the rock mass in situ, or the in-situ strength of the rock. Because of the presence of fissures, cracks, joints, gaps, stratification, discontinuities, planes of weakness, and layer anisotropy, compressibility of in-situ rock is considerably greater than that of a laboratory intact rock test specimen.

Therefore, for purposes of design in rock, it is necessary to include in equations of engineering rock mechanics the appropriate numerical values representing the corresponding in-situ rock properties.

Whereas laboratory tests render the physical and mechanical strength properties of the intact rock matrix for a relatively small specimen free from obvious structural discontinuities, and under certain casual conditions also give an idea about the behavior of rock joint, the in-situ rock tests are appropriate for evaluating the primary stress field and the mechanical behavior of the rock mass under load (30a, 31, 32, 37, 42, 47, 61, 69, 71-73, 74-78a, 80, 84, 85, 87, 91, 92).

The strength of in-situ rock depends primarily upon the extent (size, volume) of the stressed domain; e. g., the completed engineering structure may involve many and different kinds of rock weaknesses (68). However, the in-situ test of a rock would ordinarily measure the gross effect of only a few geological defects. Thus, the strength properties of a rock as obtained from laboratory tests differ from those as they prevail in situ by the following conditions:

1. In situ, the rock is confined to some extent. This means that at the commencement of work in rock, its initial stress condition differs from that prevailing in the rock specimen during its testing in the laboratory.
2. In situ, the rock mass contains various kinds of rock defects, viz., macrogeological weaknesses.
3. In situ, the rock mass is exposed to various environmental factors (climate, temperature, water) not necessarily taken into account during laboratory testing.

The in-situ data elucidate also about the mechanisms of failure of the jointed rock mass, and they render actual numerical strength values for use in solving real problems in the design and construction in rock. One should be aware, however, that because rock is a very complex material to work with, the testing of the strength of a rock mass in situ is much more difficult to perform than the testing of other homogeneous construction materials.

The basic properties of an in-situ rock mass of interest to the engineer are the mechanical properties, such as, for example:

1. Deformability (rock deformation under load, and condition of failure),
2. Compressive strength,
3. Tensile strength,
4. Shear strength,
5. Bearing capacity,
6. Permeability, and
7. Internal stresses in rock.

Internal stresses in a rock mass result mainly because of:

1. The weight of the overburden rock material,
2. Pore-water pressure,
3. Residual tectonic stresses in rocks,
4. Thermal stresses in rocks, and
5. Stresses brought about by chemical processes in the rock, mostly by recrystallization of minerals.

4-33. Deformability

Some of the factors affecting deformability are:

1. *Geological details* such as stratification, various kinds of rock defects, hydrological conditions (groundwater, pore-water pressure), and
2. *Petrographical factors* such as kind of rock mineral, degree of rock alteration, fissures brought about by excavation in rock, and internal stresses.

Introduction of an opening in the rock mass means introducing a discontinuity in the body of the rock mass, thus changing the initial or primary stress into a secondary state of stress. Hence, a change in internal stress condition is brought about.

Generally, upon loading of rock, the deformations are greatly influenced by the closing of haircracks, fissures, voids, and joints; mechanical slippage along planes of discontinuity; and inelastic deformation. Other factors influencing rock deformability are changes in pore-water pressure, temperature, drying, and vibration effects from blasting.

Up to the recent past, rock deformations were calculated by means of modulus of elasticity values as obtained from laboratory tests on rock core specimens. However, during the course of time, observations made with executed structures in rock already in service revealed that actually in-situ rock deformations were much larger than those determined from laboratory test data. The variance in these matters was explained by the existence of rock defects and discontinuities in rock. This is so because a jointed rock deforms more than a sound rock core specimen. Hence, laboratory rock specimen test results do not represent the in-situ properties of the overall rock mass.

Thus, there arose the need for testing rocks in situ. This, in its turn, creates a major problem in research, namely, how to correlate in-situ test results with those obtained in the laboratory on rock core specimens. Depending upon the extent, amount, and distribution of joints and other defects in a rock mass, the modulus of deformation of rock may be quite different from its modulus of elasticity. Recall that the modulus of deformation is based on the total measured deformation (elastic plus inelastic).

There are two methods for determining the deformability of in-situ rock, namely: the *static method* and the *dynamic method*.

In the static test, relatively large static loads are applied on the rock surface. In the dynamic test, the velocity of propagation of elastic waves is measured.

In the static test, the in-situ deformation modulus is determined by means of the so-called jack-and-plate loading test or by the pressure chamber (tunnel) test.

Some of the in-situ static tests of rocks are:

1. Plate loading (bearing),
2. Jacking,
3. Pressure chamber (tunnel),
4. Compression,
5. Bore-hole deformation,
6. Tension,
7. Shear strength,
8. Torsion, and
9. Stressing of the rock in the interior of a bore hole by the so-called method of hydrofracturing of the rock.

4-34. In-Situ Testing of Rocks

The main *objectives* of in-situ testing of rocks are:

1. To assess the general suitability of a site for contemplating engineering work,
2. To facilitate adequate, safe, and economical design of geotechnical structures for that site,
3. To foresee and to provide against geotechnical problems during and after construction, and
4. To investigate any subsequent changes in site conditions, or any eventual failures which may occur during construction.

Plate Loading Test

The purpose of the plate loading test is to determine the extent to which the rock mass will deform under various externally applied loads. Sometimes it is possible to increase the loading until failure of a weak rock material comprised of blocks and slabs of rock.

In this method of test, a normal load is applied to an exposed, flat surface of the rock by means of a hydraulic jack, and the resulting displacement of the rock surface is measured.

The theoretical basis for this test is the well-known, modified Boussinesq elasticity solution (19) for the distribution of stress and strain in a semi-infinite elastic medium subject to a surface load N:

$$w_o = \frac{m \cdot (1 - \mu^2) \cdot N}{E \cdot \sqrt{A}} \tag{4-106}$$

where w_0 = surface displacement of rock, viz., average deflection of a rectangular loading plate brought about by a total distributed load N on the plate;

m = displacement coefficient;

μ = POISSON's ratio;

E = YOUNG's modulus of elasticity;

A = area of the loading plate; and

N = total normal surface load.

According to Reference 89, the displacement coefficient values of m for circular, square, and rectangular loading plates are those given in Table 4-12.

TABLE 4-12

Values of *m*

Form of loading plate	Side ratio a/b	m values	Reference
(1)	(2)	(3)	(4)
Circle	—	0.96	89
Square	1 : 1	0.95	
Rectangle	1 : 2	0.92	
	1 : 5	0.82	
	1 : 10	0.71	
	1 : 100	0.37	

Usually three to six plate loading tests are made at each location. Individual tests are usually separated by five plate diameters.

In each test, loads are usually cycled, and a secant modulus will usually be taken from the hysteresis loops at the stress level of design interest. Each test usually consists of four cycles of loading and unloading, with progressively higher loads and recording of the rock deflections. For example, in the first loading cycle, 25 % of the full load is applied; 50 % in the second cycle; and so on. A typical cyclic deformation graph from repeated loading is shown in Fig. 4-38.

In some rocks, when the load is cycled several times, the inelastic deformation increases.

Besides simply applying an increased load increment on the plate, it is also necessary to control the rate of loading because it, too, can affect the mechanical properties of a rock.

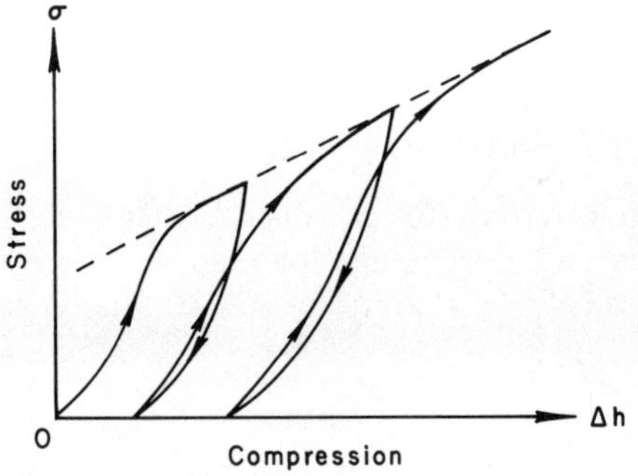

Fig. 4-38
Cyclic loading and unloading diagram

Small-scale in-situ tests confined to a relatively small volume of rock may give misleading results. Therefore, for large projects the tendency for in-situ testing is toward larger, comprehensive volumes of rock.

Depending upon the nature of the rock material tested, failure of rock can be recognized by a definite fracture, by a gradual continuous yielding, or by a sudden yielding of the rock.

Rock in situ may rupture as a result of:

1. Slip along planes of discontinuity,
2. Lack of cohesion, and
3. Plastic failure.

The physical constants used in estimating the load under which a rock mass in situ might fail are:

1. Insufficient angle of friction φ,
2. Insufficient cohesion c, and
3. The level of the stress applied to the rock, viz., exhaustion of its shear strength.

The plate loading test on rock provides a convenient means for determining the modulus of deformation as well as the in-situ strength of the rock.

Jacking Tests

Deformability of an in-situ mass of rock is usually determined by jacking (or jack-loading) tests (31, 42, 48, 69, 72, 84) the principle of which is shown in Fig. 4-39. The test is usually carried out in excavated underground test galleries in rock (2, 21). Here the jacks are supported against the rock ceiling and floor of an underground opening. The load is applied by means of a jack placed on a flat surface of the rock, and the resulting deformation is measured. The data so obtained are utilized in computing the modulus of elasticity, viz., deformation modulus of the rock. The bearing plate in this test must be large enough to comprehend the structural features which contribute to the characteristics of the rock mass.

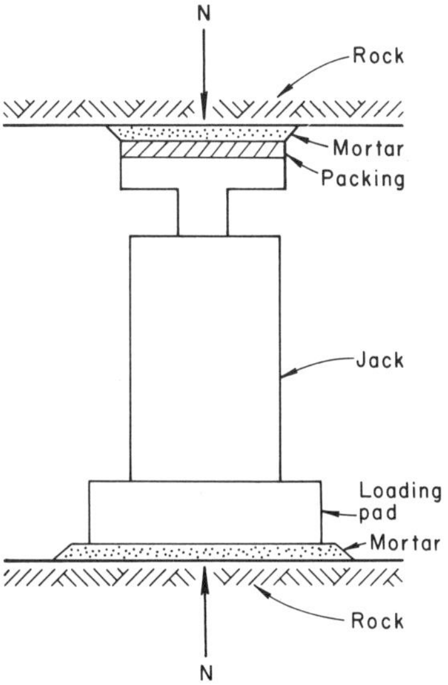

Fig. 4-39
Tunnel jacking test

Jack-loading tests have two serious disadvantages. First, no matter how carefully the excavation of the underground opening (tunnel, gallery, adit) or trench for the tests are made, excavation inevitably disturbs the rock mass surrounding the excavation. The second disadvantage is the necessity to limit the loaded area. This, in its turn, decreases the representativeness of the results and decreases the number of tested spots in the rock mass.

Cable Jacking Test

For performing plate bearing tests on flat, open surfaces of rock, STAGG (1968, Ref. 84) proposed that the jack reaction be taken up by cables anchored at suitable depth into predrilled bore holes in rock below the load-bearing plate (Fig. 4-40). It seems that this type of test would be more economical to perform than the simple plate loading test. Another advantage of the cable jacking test is that the rock can be tested at the exact location of the foundation, as well as in the directions in which the structural loads will actually be acting.

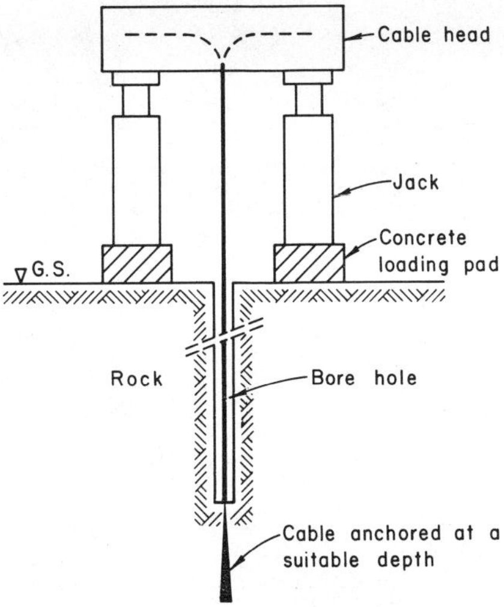

Fig. 4-40
Principle of cable jacking test

Pressure Chamber Test

The pressure chamber test (42, 47, 72, 73, 80, 84), known also by the term pressure tunnel test, consists essentially of sealing off a part of a tunnel by means of bulkheads. Prior to that, to avoid loss of water from tunnel to rock, a concrete lining is placed over the test tunnel surface of the rock, and the inside surface of the tunnel lining is coated with a waterproofing material for water tightness. Then the instruments are affixed for strain measurements. The chamber so prepared is then hydraulically pressurized, water usually being the pressurizing medium. Then the rock deflections brought about by the internal water pressure in the tunnel are remotely recorded at stations outside the sealed-off part of the tunnel. Thus in this kind of test the rock wall of the circular tunnel is subjected to a uniformly distributed radial loading, bringing about diametral deformation Δd. This diametral deformation is used to calculate the YOUNG's modulus of elasticity E:

$$\Delta d = 2 \cdot \frac{(1 + \mu)}{E} \cdot r_i \cdot p \tag{4-107}$$

and

$$E = \frac{(1 + \mu)}{\Delta d} \cdot d \cdot p \tag{4-108}$$

Herein, Δd = deformation of the rock measured behind the skin of the chamber;

μ = POISSON's ratio, usually obtained from laboratory tests on rock core specimens (it can also be obtained by geophysical test techniques by measuring the velocity of the shear wave within a rock mass);

r_i = $d/2$ = initial radius of tunnel;

d = diameter of tunnel; and

p = intensity of uniformly distributed radial, internal, hydrostatic pressure.

Because of the anisotropic nature of the rock, the E-values are to be determined for more than one direction within the rock mass.

From these measurements, pressure versus deformation diagrams are plotted, and the YOUNG's modulus of elasticity may be obtained.

One important feature of the pressure tunnel test method is that it brings about tensile hoop stresses in the rock. If this stress exceeds any residual compression of the rock, it may open radial cracks in the rock.

Some of the advantages of the pressure tunnel test are:

1. Rock deformations (deflections) can be measured in any radial direction.
2. Hydrostatic pressure on rock in the tunnel can be sustained for a long period of time.
3. The effect of the concrete lining on deformation modulus can be determined by testing lined and unlined rock in the same chamber.

The main disadvantage of this kind of in-situ test is that it is relatively expensive.

Large-Scale Compression Strength Test

The rock mass compressive strength is determined from tests on rock pillars. Their strength is a function of the rock material, the geometry of the pillar, its size, and the compressive strength of the rock strata between which the pillar is compressed.

Borehole Deformation Test

Deformability of rock in situ is also possible to test in boreholes by means of a device called the borehole deformation meter (33, 37, 38). In principle, the device consists of a shell split into two halves. It is inserted and lowered into the borehole. Then the shell is filled with oil and pressure is applied on the oil. Thereupon the split shell bears laterally against the rock, thus imparting pressure on the rock wall of the borehole. The principle of the working of a borehole deformation meter is sketched in Fig. 4-41. A sensitive transducer is used to measure the deformation of the rock wall

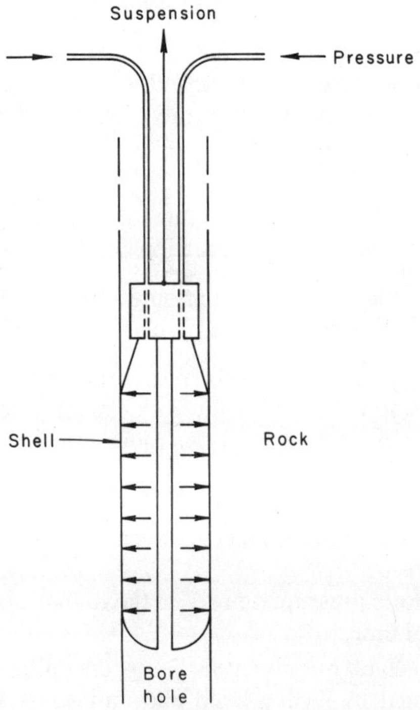

Fig. 4-41
Principle of the working of a borehole deformation meter

of the borehole. Various loading and unloading cycles can be performed by means of this meter, and the in-situ modulus of deformation can be determined.

In-Situ Tension Test

Any incompetent rock mass in its *natural state* cannot sustain tensile stresses because of the presence of joints, fissures, and other kinds of rock defects. Therefore, from a *structural* point of view, the tensile strength test in situ is not of too great importance, except in a layered roof of an underground opening where tensile stresses of considerable magnitude may have to be coped with. Tensile stresses in rock may also be of importance in the stability of rock slopes over pre-existing planar discontinuities, where the dipping rock layers may be subjected to tensile stresses and thus slope failures would be more probable.

4-35. In-Situ Shear Strength Test

The main purpose for performing in-situ rock shear strength tests is to obtain shear strength values that comprehend the effect of the various rock defects (2, 3).

The test area should be chosen to represent the area on which the structure will be built.

To test the shear strength of the in-situ rock mass, the rock specimen to use in such a test should be as large as possible. The size of the specimen should not be less than 40 cm square by 20 cm high. If the specimen is larger than 40 cm x 40 cm x 20 cm, the ratio of length, width, and height is usually taken as 2:2:1. Sometimes the size of the base area of the rock block is taken as 0.70 m x 0.70 m, or even as 1.0 m x 1.0 m. The shear test should be made on rock that can be trimmed without damage.

The in-situ shear strength tests are usually performed on rock blocks cut in the rock mass but kept attached to it in their bases. After the block has been properly prepared, the normal and tangential loads are applied to the block by means of hydraulic jacks (Fig. 4-42). In an underground opening, the jacks are supported against the ceiling and wall of rock.

The vertical jack supplies the normal load N on the forced shear plane s-s of the rock, whereas the second jack (horizontally installed, or installed at an inclination with the horizontal) supplies the shear load.

The shear thrust may be vectored at an angle of about $(\pi/4 - \varphi/2)$-degrees to the position of the shear plane. Thus an isolated block of rock is sheared off from the parent rock. During loading and unloading, measurements of deformations are recorded.

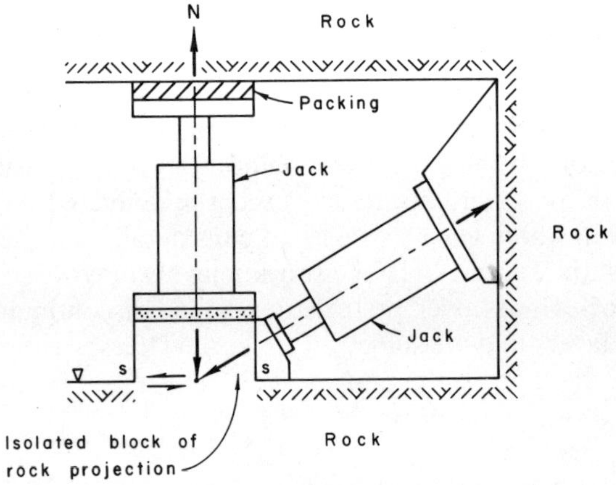

Fig. 4-42
In-situ shear strength test in an underground opening

This kind of in-situ shear strength test gives an idea of the magnitude of the angle φ of the shear resistance of the rock. It is assumed here that COULOMB's shear strength law applies:

$$s = \sigma_n \cdot \tan \varphi + c \tag{4-109}$$

wherein s = shear strength,
σ_n = normal stress on the shear plane,
φ = angle of shear resistance of rock, and
c = cohesion of rock.

Because the shear strength of rock is given by two shearing strength parameters, φ and c, two equations are needed for their determination. Therefore tests are made with bearing plates or blocks of various sizes.

Depending upon the need in design, shear strength tests in situ may also be performed on oriented planes of separation or geological discontinuities such as planes of bedding, shistosity, jointing, and faults. In these planes of weakness, the least shear strength of a rock mass can thus be anticipated.

The in-situ shear strength of rock in an underground opening may also be determined as sketched in Fig. 4-43. This kind of a test arrangement has the advantage that both normal and shear stresses are applied on the ultimate plane of shear failure s-s using only one jack.

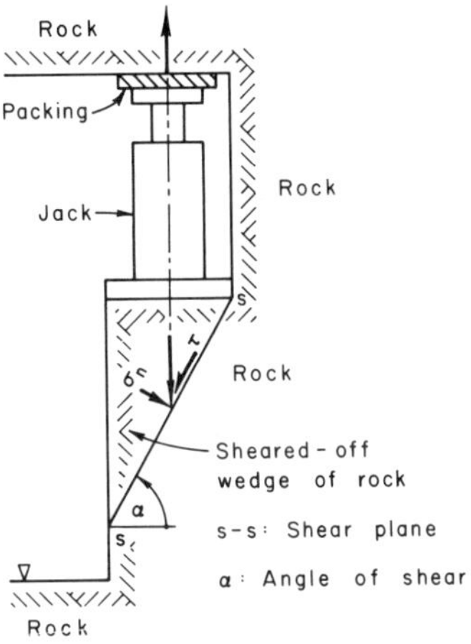

Fig. 4-43
Shear strength testing of rock in situ with one jack

The principle of in-situ shear and in-situ sliding tests is illustrated in Fig. 4-44. Generally, under small normal loads, the rock blocks shear off over a plane contact surface (Figs. 4-44, 1 to 6).

However, in evaluating heavily loaded (*N*), large-scale, direct, in-situ shear test results on rocks, one should be aware that no shear takes place along the contact plane between the block and rock, but rather over a gently curved sliding surface in the rock on the leeward side of the loaded block. This pertains also to testing of concrete blocks on rock (Fig. 4-44, 7). The same phenomenon has also been observed in the author's experiments with obliquely loaded, rigid foundation models on dry sand at relatively light, normal contact pressures (Fig. 4-45), Refs. (50-52).

For repeated tests, several shear blocks of rock are necessary.

Because of the irregular distribution of the various kinds of rock defects, the determination of the in-situ shear strength of a rock mass is a difficult task indeed.

Frequently, the sliding surfaces in a rock mass are not smooth, but are irregular, rough, and wavy. Often the sliding surfaces are filled with gouge (for example, geological debris such as silt, clay, sand, ground rock flour, calcite, and other kinds of decomposed rock material). Depending upon the degree of waviness, joints during a shearing, viz., sliding process, may open and/or close (Fig. 4-46). A relatively thin course of clay in the plane

Symbol of shearing system	Kind of in situ shear test	Test results
① 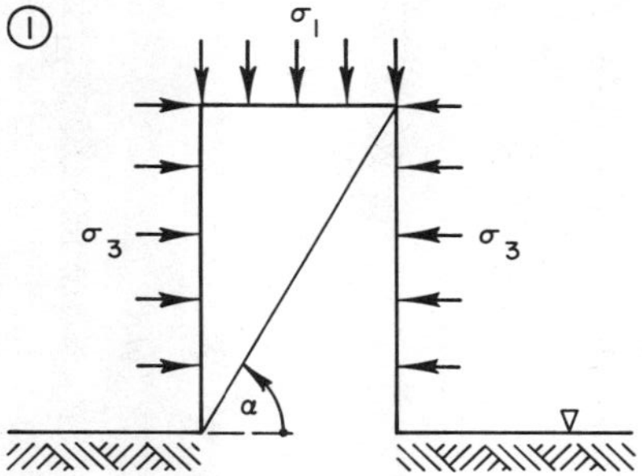	Compression confined $(\sigma_3 = \sigma_3)$ unconfined $(\sigma_3 = 0)$	Compressive strength. Deformation characteristics. Stress – strain diagram. Elastic properties. Shear strength parameters ϕ and c.
② 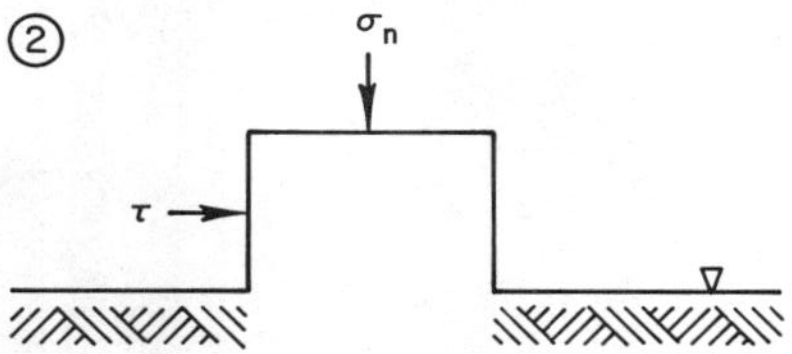	Shear homogenious material	Shear displacement. Shear strength. Parameters ϕ and c.
③ 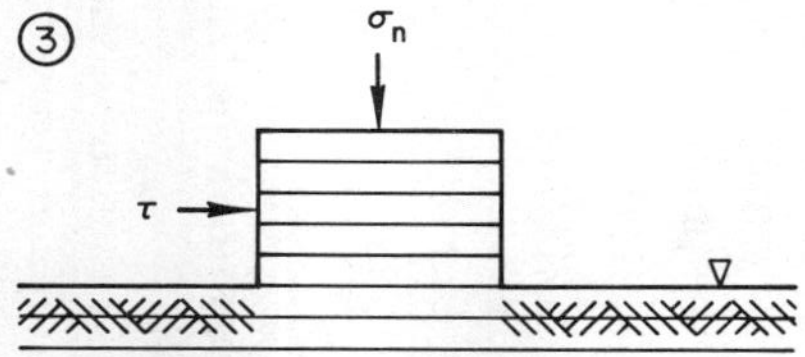	Shear parallel to bedding planes or laminations with σ_n; without σ_n	As under ②, but along bedding planes.

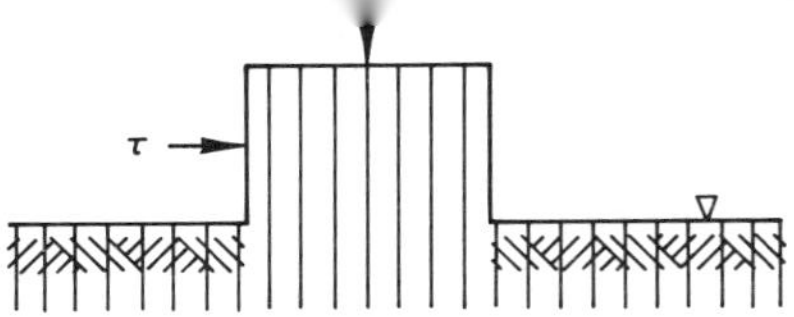	bedding planes or laminations	perpendicular to bedding planes.
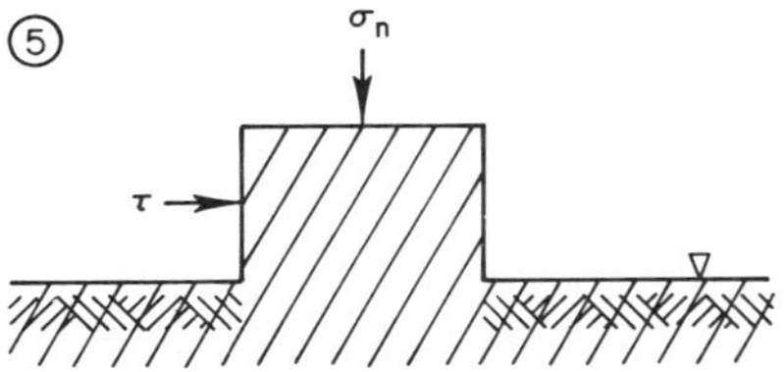	Shear of dipping beds or laminations	As under (2), but for dipping system of layers of rock.
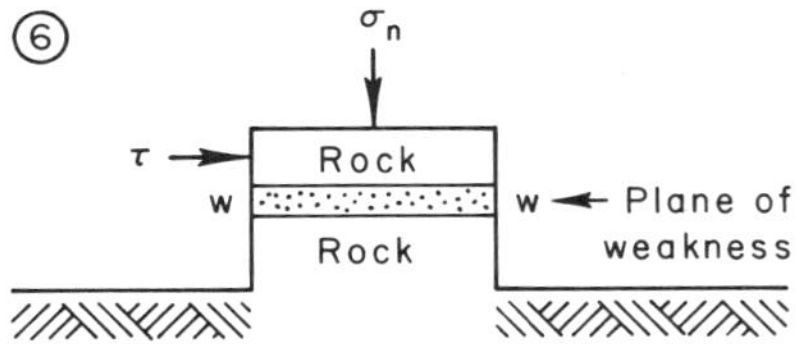	Shear along seams and of gouge	Frictional resistance along an existing plane of weakness of a rock defect. Shear strength of gouge; ϕ and c.
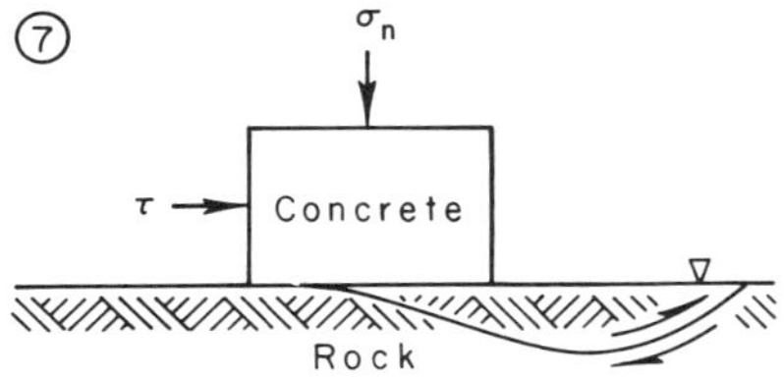	Shear between concrete and rock Shear of rock	Deformations. Bonding strength between concrete and rock.

Fig. 4-44
Some principles of large-scale shear strength tests

Fig. 4-45
Rupture surface pattern in dry sand brought about by an obliquely, centrally loaded foundation model at a contact pressure of $\sigma_0 = 1.0\ kg_f/cm^2$ (Author's study, Refs. 50—52)

of weakness or discontinuity in rock, and fillings in the joints between beds and rock blocks, may considerably alter natural stress conditions in the rock mass, as well as bring about failures of stability such as those which were experienced at Malpaset (45) and Vaiont dams (59, 63, 66, 67, 82); the Madison Canyon (17), and at other sites.

Also, pore-water pressure affects sliding of rock masses on and along faults, joints, and bedding planes, thus impairing the stability of a layered rock structure. Thus, joints, shear zones, and other kinds of rock defects in a rock mass decrease the magnitude of the shear strength value much below the shear strength of the intact rock specimen of the same rock. Therefore, the in-situ shear strength of a rock is to be qualified as very anisotropic. Relative to design and construction in rock, this means that the most critical situation is one when the shear strength of the rock mass is governed by the shear strength, or resistance along the rock contact surfaces of a discontinuity. Therefore, rock defects of a rock mass generally cause scatter in the results of the shear strength test values. The difference in test results decreases with increase in size of the test area. Also, the in-situ strength increases with decrease in the amount of discontinuities (jointing).

Values of the shear stress parameter φ_r of residual shear strengths obtained from most types of rock, exclusive of clay shales, usually lie between $\varphi_r = 25°$ and $35°$, while the φ_r-values for most clay shales lie between $5°$ and $25°$ (74).

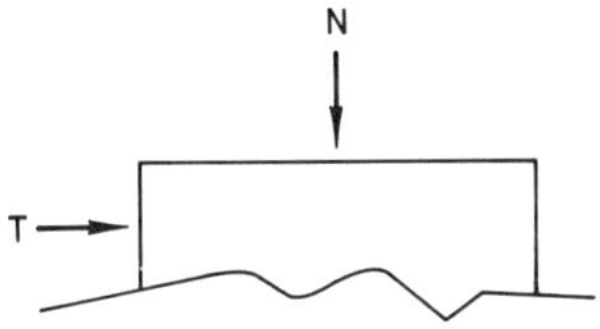

a) Shearing along planes of separation

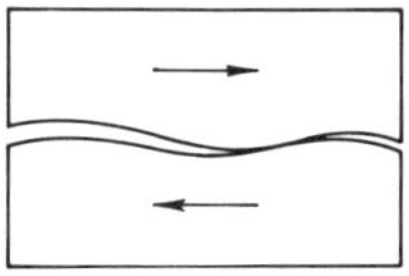

b) Opening of a joint upon sliding

Fig. 4-46
Rough and wavy sliding surfaces

Torsion Shear Test

The method of torsional shear test involves the application of a torque to a cylindrical rock core which is isolated without damage from the surrounding rock mass by means of a diamond rock core drill. The rock core so prepared is kept attached at its base to the rock mass (Fig. 4-47). The core is then gripped by the testing device and twisted. The angle of twist and the failure are measured and recorded. The test renders information about the test parameters c (cohesion), φ (angle of friction), and residual shear strength of the rock.

The torsion shear test is the only method of bringing about shear in a rock specimen without compression.

In order to set up a perfect torsion shear test, the influence of clamping or gripping effects and of all eventual rock core bending effects on test results should be eliminated.

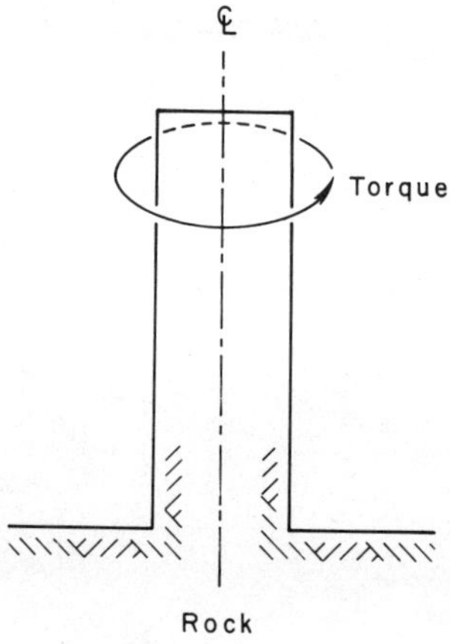

Fig. 4-47
Principle of in-situ torsion shear test of rock

4-36. Internal Stresses in a Rock Mass

The prerequisite for safe and effective design and work in rock is a comprehensive knowledge about the physical and mechanical properties of in-situ rock. These properties, and rock anisotropy, are known to vary with depth of the rock formation. The anisotropy of an in-situ rock mass may be evaluated by knowing its natural stress condition.

The main uses of rock stresses are in:

1. Designing of tunnel linings;
2. Designing of walls, pillars, roof arches, and domes for underground openings in rock;
3. Evaluation of rock stability of various kinds of structures in rock, including stability of rock slopes made in open cuts; and
4. Evaluation of stability of rock-abutment systems of high dams.

Internal, natural stresses in a rock mass in situ result mainly because of:

1. The weight of the overburden rock material;
2. Hydrostatic pressure;
3. Pore-water pressure;
4. Thermal stresses in the rock;
5. Residual tectonic stresses in the rock; and

6. Possible stresses brought about by chemical processes in the rock, mostly by recrystallization of rock minerals.

Factors affecting natural stresses in rock are:

1. Local stresses brought about by excavation in rock,
2. Effects of blasting,
3. Seismic forces,
4. Changes in hydrostatic pressure,
5. Changes in pore-water pressure,
6. Changes in rock thermal regimen, and
7. Variation and alteration in physical and chemical factors of the rock mass and its minerals.

In-Situ Test of Stresses in a Rock Mass

Any underground opening in rock is deformed by stresses in it (31). The opening deforms upon change in stress in the rock. However, it must be recognized that the concept of stress is a theoretical one, and that it is difficult to measure stress directly. In the same vein, therefore, it must be said that presently there is no way of measuring natural stresses in the interior of a rock mass without disturbing the rock at the location of measurement. However, in order to get a value for the stress at a point, other properties of the rock relating to stress are usually measured and calibrated. Thus, one resorts to measuring the stress by indirect methods. In these methods, usually elasticity of rock is assumed, strains in rock are measured, and strains and stresses are then related. The underground measurements further the understanding of the performance of rock around an underground opening, and provide for basic design data. Thus, the information about the stress field in the rock is derived from displacement, viz., strain, measurement, which can be used to measure changes in stress in rock around the opening. Hence, the analysis of strain is fundamental in the studies of deformation of the rock.

Some of the methods of measuring strains that lead to state of stress determination in rock are:

1. Direct strain measurement,
2. Dilatometer test,
3. Overcoring technique,
4. Flat-jack testing,
5. Electrical resistivity,
6. Seismic wave velocity, and
7. Photoelasticity analysis of stress.

Direct Strain Measurement

In this method, strain gauges are cemented on an exposed rock surface, or in a borehole, and the rock strains are measured.

Dilatometer Test

Deformation of in-situ rock mass can also be determined along the inside of boreholes by means of instruments called borehole deformation meters and dilatometers.

The dilatometer, developed in Portugal (ROCHA, 1966, Ref. 18), imparts a uniform pressure along a certain length (depth) of the borehole by means of a liquid contained between a metallic cylinder and a very deformable rubber or steel jacket. The jacket is applied against the rock wall of the borehole.

The diametrical deformation Δd of the diameter $d = 2r$ of the borehole calculates as:

$$\Delta d = 2 \frac{1 + \mu}{E} \cdot r_i \cdot p \tag{4-110}$$

From here, the modulus of elasticity E of the rock is:

$$E = 2 \frac{1 + \mu}{\Delta d} \cdot r_i \cdot p \tag{4-111}$$

where μ = POISSON's ratio,
r_i = inside radius of the borehole, and
p = intensity of dilatometer pressure.

The pressure p is measured by means of a pressure gauge, and the diametrical deformation of the borehole, viz., strains, is measured along the external face of the dilatometer by means of transducers.

Compared with the load test, measurements made by means of the dilatometer inside boreholes have the following advantages:

1. If diamond drilling is used, the rock mass remains almost undisturbed.
2. Because of the reduced cost and time required in performing dilatometer tests, many tests can be performed. Thus it is possible to obtain pertinent statistical data about deformability of an anisotropic rock mass.
3. Dilatometer tests can be performed at great depth.

4. Such tests can be performed under submerged condition.
5. Much higher pressures can be applied on the rock than in the load test.

A disadvantage of the dilatometer test is the small volume of rock involved in each such test.

Whereas a domain subjected to a jack-loading test is free from or decompressed of internal or initial stresses in the direction of the applied load, the volumes of rock involved in the dilatometer tests are in a state of radial compression (zero at the surface of contour of the hole) when the rock mass exhibits radial stresses. Therefore it should be expected that the deformability determined by means of dilatometers lies below the values obtained in jack-loading tests, because usually deformability decreases with increase in compressive stresses.

There have been several types of dilatometers developed in various countries. Those instruments in which a radial, uniform pressure is applied by means of a liquid on the rock wall of the borehole appear to be the best ones for measuring deformability in a rock mass. The other instruments induce more involved states of stress in the rock mass, thus making the interpretation of the test results somewhat difficult.

Flat-Jack Testing Method

The principle of in-situ state of stress measurement in a rock mass is: the relief of stress, and its recovery by means of hydraulic pressure. Such a test is accomplished by means of a flat jack. Essentially, the flat-jack testing technique may be described as set forth (78a).

A narrow slot is cut into the rock wall, such as the wall of a tunnel, for example (Fig. 4-48). Prior to cutting of the slot, two displacement or deformation measuring pins are fixed into holes drilled into the rock, or cemented onto the surface of the rock wall. Upon cutting the slot, the stresses in the rock originally across the surfaces of the slot are relieved or relaxed on either side of the surfaces of the slot. Upon this stress relief, expansion of the rock into the slot takes place.

The rock expansion or displacement can be precision-measured either by the convergence of the rock across the slot, or by the convergence between the measuring pins on opposite sides of the slot. After the slot is cut, a thin flat jack is placed into the slot cut into the rock wall. The flat jack is filled with a fluid, usually oil, and pressure in the jack is applied. The pressure is increased until the displacements or deformations brought about by cutting the slot are cancelled (reversed, or until the measuring points return to their original position). This means that in such a case the flat-jack pressure p attained the value of the original normal stress σ_z of the rock initially existing in it before cutting the slot; i.e., $p = \sigma_z$. This pressure p, thus, represents the stress acting in the direction normal to the flat jack. Thus, in a way, the flat-jack testing method may be regarded as a null technique by means of which rock stresses can be measured directly.

Some of the advantages of the flat-jack testing method are:

1. The method is very simple and relatively effective in sound, hard rock.
2. Disturbance in the rock mass is reduced to a minimum.
3. It renders vertical and horizontal absolute stresses around an exposed face of rock.
4. Slots in rock can be cut, and flat-jack measurements can be made at any angle with the horizontal.

Some of the disadvantages of the flat-jack testing method for determining the state of stress within a rock mass at its surface in an underground opening are:

1. The initial state of stress in rock is changed already by excavation.
2. The method is limited to measuring of stresses near the exposed surface of the rock wall.
3. It is complicated by the effect of creep of the rock during the period between cutting of the slot and stress cancellation.

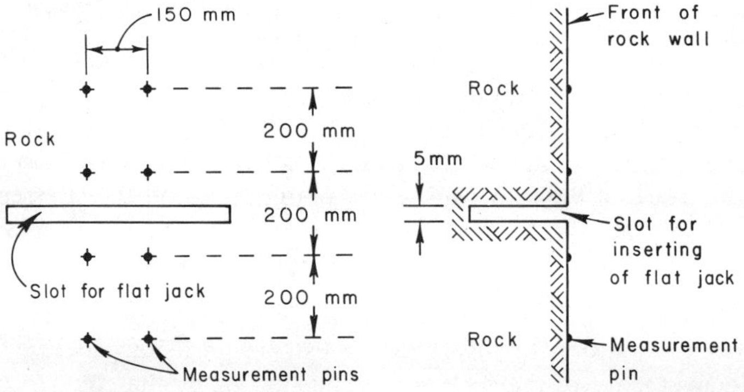

a) Front view of slot in rock

b) Section through slot in rock wall

Fig. 4-48
Slot in a wall of rock for performing a flat-jack test
(After ROCHA et al., Ref. 78a)

Overcoring Technique

The overcoring technique is also known by the term borehole strain measuring technique. This technique is used to determine the "absolute stress" in the rock, according to LEEMAN and HAYES (1966, Ref. 61).

In essence, the technique involves measuring strains at three points around the periphery of a borehole made in the "roof", or in the "sidewall", or at a point midway between the "roof" and the "sidewall". By using strain gauge rosettes of suitable configuration, sufficient strain measurements in three different planes are, in effect, then obtained.

This method of absolute stress measurement in situ may be described briefly as follows. An NX borehole is drilled into rock to a depth at which the stress is to be determined. Then an EX borehole is concentrically drilled for, say, 45 cm into the end of the NX borehole. Then three rosette strain gauges are glued around the periphery of the test borehole by means of a special tool in the EX portion of the borehole made in the rock, and spaced in positions at $\theta = 0°$ (at the roof of the borehole), at $\theta = \pi/2$ to the first, and at $\theta = 5\pi/4$ to the first rosette. Then strain readings are taken. After that, the EX portion of the borehole is overcored with an annulus (Fig. 4-49), using an NX casing crown. Then the hollow cylindrical core containing the rosettes is removed and the stress-relieved strain readings are taken from each gauge. Each strain gauge rosette has three gauges to measure strains in the circumferential, axial, and 45-degree directions.

The standard sizes for casings, rods, core barrels and holes are compiled in Table 4-13.

Table 4-13

Standard Sizes, in inches, for Casings, Rods, Core Barrels and Holes

Casing Core Barrel	Drill rod	Casing OD	Core barrel bit OD	~ Diameter of bore hole	~ Diameter of rock core specimen
EX	E	1-13/16	1-7/16	1-1/2	7/8
AX	A	2-1/4	1-27/32	1-7/8	1-3/16
BX	B	2-7/8	2-5/16	2-3/8	1-5/8
NX	N	3-1/2	2-15/16	3	2-1/8

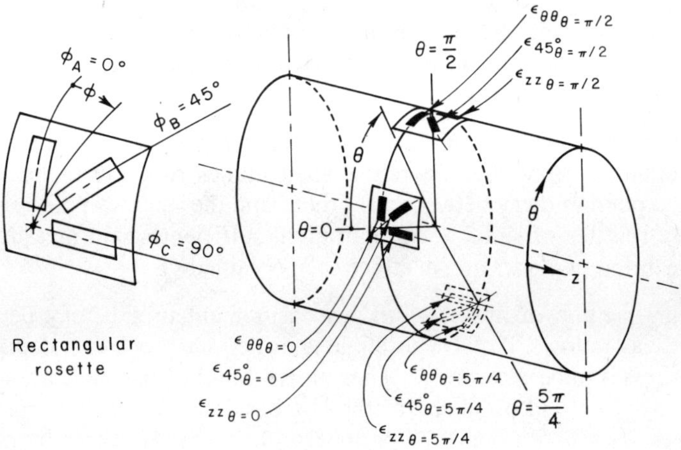

A. Positioning of strain gauge rosettes
ε = strain

Fig. 4-49
Overcoring technique
(After LEEMAN and HAYES, Ref. 61)

It is advisable to use as large a diameter of core as economy will permit. The larger the diameter, the more accurate are the geological observations on jointing and fracturing. Very small diameters such as EX are to be avoided when possible: many widely spaced fractures may be missed entirely.

The difference in the strain readings before and after overcoring enables one to determine the complete stress in the rock, provided that the properties of the rock are known.

The removed core can be inspected in the laboratory for bonding condition of the strain gauge rosettes.

In this method, it is easy to measure the changes in strain. The method is a rapid one of obtaining absolute strain. Its accuracy, however, is open to question, particularly over a period of time.

According to LEEMAN and HAYES (1966), the term "complete state of stress" at a point in rock means the magnitudes and directions of the three principal stresses acting at the point (61).

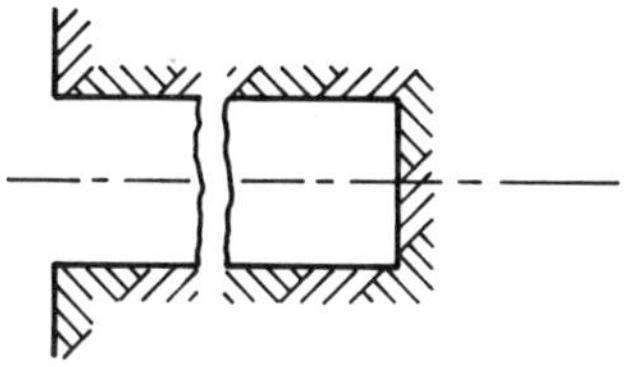

(a) NX Borehole drilled to the depth at which the stress is to be determined

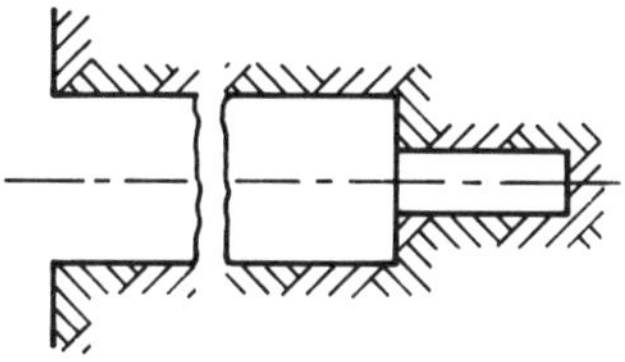

(b) EX Borehole drilled for 18 ins. (say) into the end of the borehole

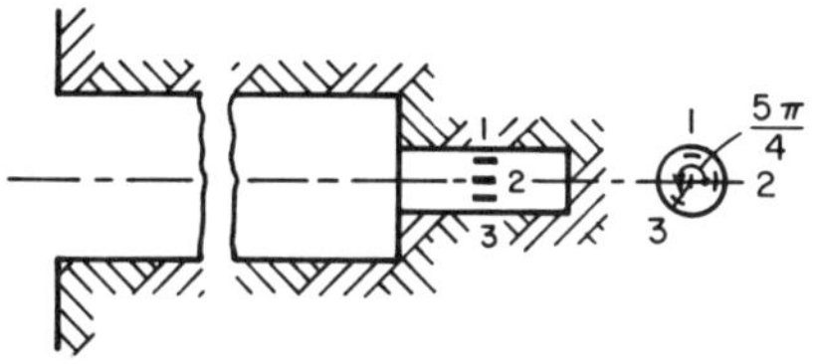

(c) Three rosette gauges glued in the EX portion of the borehole and strain readings taken

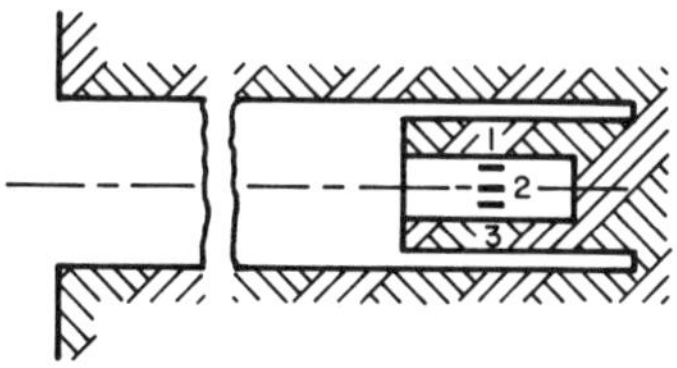

(d) EX portion of the borehole overcored using an NX coring crown

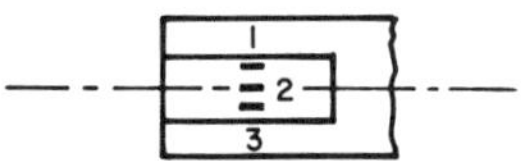

(e) Cylindrical core removed and strain readings taken

B. Sequence of overcoring

Fig. 4-49 (Continued)
Overcoring technique
(After LEEMAN and HAYES, Ref. 61)

Hydraulic Fracturing Test of Rock

Hydraulic fracturing of a borehole is used extensively in the oil and natural gas industry to stimulate yield from a well. The method of hydraulic fracturing of rock is also used in rock engineering.

In principle, the method of hydraulic fracturing of deep borehole rock walls should render the horizontal, tectonic, tangential, principal stress σ_t. In this test, a certain section of the borehole is sealed off with packers such that a fluid pressure can be applied to the bare wall of rock of the borehole between the packers only to a restriced range of depth of the hole, viz., thickness of a layer of rock. Then water (or any other liquid) is pumped into the borehole for fixed periods of time to a sufficient and successively higher pressure. This pressure p is increased until fracture of rock sets in the wall of the rock. This means that the tangential stress σ_t in the surface of the contour or periphery of the borehole becomes tensile. This tensile stress brings about fracture in the rock if this σ_t stress exceeds the tensile strength σ_{t0} of the rock. In other words, a crack intersecting the borehole — which crack is kept "closed" by the initial compressive stress field in the rock — may become "opened" by the newly induced tensile stress after a hydraulic pressure p is applied.

As long as the flow of the fluid being pumped is laminar, the diagram of the amount of water acceptance Q by the rock plotted against the applied pressure p will be a straight line. Any sudden increase in the slope of this (Q-p)-diagram may indicate that rock fracturing takes place.

The cracks in the rock extend or propagate as the pressurized liquid flows into the cracks.

In rock mechanics, it is believed that the cracks propagate normal to the direction of the minor principal stress. Also, the pressure p of the liquid bringing about propagation of cracks may be considered as the magnitude of the minor principal stress in the rock. The pressure p for cracking the rock is somewhat larger than the magnitude of the minor principal stress.

Referring to the pertinent literature (Fairhurst, 1964, Ref. 30a) it must be said, though, that the method of hydraulic fracturing of rock as a means of its use for determining horizontal tectonic principal stresses in rock has aroused a great deal of discussion.

4-37. Permeability

Permeability is defined as that property of a porous material that permits the passage or seepage of fluids — such as water, or oil, or gas (air) — through its interconnecting voids.

The factor permeability of rock to water in rock masses is also of paramount importance, not to be forgotten in designing safe structures in rock. The most frequently observed flow in rock is the flow of water through joints and other rock defects. It should be remembered that the presence and flow of interstitial water (in pores, along faults, in fissures, in crushed zones, and in gouges) frequently bring about hydrostatic uplift and pore-water pressures. These pressures affect seriously the stability of foundations and rock abutments of dams and the roofs of underground openings, and may alter the internal stress condition in the rock. Knowledge about permeability of rock is also needed for executing injection and/or grouting projects in rock. More about permeability of rock to water may be found in Sections 4-4, 4-8, and 10-3, and in Refs. 44 and 80.

The flow of water in rock in situ is determined by means of permeability tests.

Permeability tests of a rock mass in situ are usually made in a system of several boreholes by measuring loss of water to joints. When joints are closed and filled, and the rock is permeable, permeability tests render diagrams similar to those obtained for soils.

There are two kinds of borehole permeability tests:

1. Bottom-hole method, and
2. Packer method.

The bottom-hole permeability test consists of simply blocking one end of the hole and filling it with water. The time required for the water level to fall a given distance is then recorded.

In the packer test, several test holes, ~ 5 cm in diameter and at various depths, are drilled into the rock. These boreholes are sealed with rubber packers; then the holes so sealed are filled with water. The flow of water, the maximum pressure, and pressure changes in each hole are recorded.

Full-scale water permeability tests can also be made in aqueous tunnels. Such tests are supposed to indicate loss of water from a rock opening in the case of an aqueous tunnel, and the amount of ingress of water from the rock into the underground opening in the case of a vehicular tunnel, underground power plant, and other kinds of openings.

Permeability test results are also used in evaluating the feasibility of constructing a lined rather than an unlined tunnel.

4-38. Electrical Resistivity of Rocks

Rock electrical resistivity data may be used for analyzing stresses in rock. However, the data obtained in this test are difficult to interpret because the electrical resistivity of rock is affected by the kind of mineral content of rock, its porosity, moisture content, and connecting of the electrodes of the various instruments involved, as well as the difficulty of calibration for tectonic stress conditions.

CHAPTER 5

ROCK MASS PROPERTIES

5-1. Mechanical Defects of Rocks

All rock mass properties may be grouped into two large groups, namely:

1. Macroscale, and
2. Microscale.

Macroscale rock mass properties are essentially properties of the component.

Microscale rock properties are properties of the substance. In evaluating static and dynamic conditions within the body of the rock *en masse*, engineers consider natural inferiorities or weaknesses (mechanical defects) of rocks. The inferior properties of a rock mass affect underground construction operations in rock, and performance of structures built on or in rock.

Without exception, large masses of all kinds of rocks, regardless of the type, have more or less conspicuous mechanical or structural defects or natural inferiorities that influence adversely the strength of rock *en masse* and engineering operations in rock. These defects consist of more or less closely spaced fractures and other *planes of weakness* in rock. The existing weaknesses may be naturally produced systems of more or less closely spaced fractures in rock *in existence,* such as

1. Fractures, cracks and haircracks;
2. Fissures;
3. Bedding planes, laminations, schistosity, partings;
4. Stratification;
5. Joints;

6. Fault planes and zones, crushed zones;
7. Folds;
8. Voids;
9. Cavities;
10. Seams and interbeds of weak and plastically unstable rocks, aquifers, clays, and shales; and
11. Ancient slip planes and other possible weaknesses.

Planes and zones of weakness in rocks are classified as existing and/or potential ones.

Bedding Planes

Bedding planes are the division planes in bedded deposits such as sedimentary and stratified rocks, which planes separate the individual layers, beds, or strata.

Cleavage Planes

Cleavage is the tendency to split or cleave along definite, smooth, parallel, closely spaced planes. As applied to rocks, cleavage is the property of splitting into thin, parallel sheets. The cleavage plane is the plane in rocks or minerals along which cleavage takes place.

A *parting* is a thin layer of deposited and altered material (carbonaceous or other organic material) separating beds in sedimentary or metamorphic rocks.

Separation is a relatively fresh break along a bedding plane or between beds in sedimentary or metamorphic rocks. Separations are usually man-made: they develop as a consequence of mining.

Schistose rock: Any crystalline rock whose constituent minerals have a more or less parallel arrangement.

If the walls of the joint are separate, the term *fissure* is sometimes used.

If a rock has innate mechanical defects such as bedding or cleavage planes, the joints and faults constitute an additional source of weakness.

Potential planes and zones of weakness in rock are artificially created weaknesses in rock sequences. These may be brought about by man's activities during broad-scale, intensive blasting operations in rock; the possibility of influx of water in and seepage through fissures, cracks, joints, and fault zones in rock; and possibly by other factors.

Naturally and/or artificially produced planes and zones of weakness in rocks affect the stability of underground openings and the stability of excavation slopes made in rocks, and provide for passageways for the underground flow of water.

Investigations of planes and zones of weakness caused by man's activities have been pursued very meagerly; they have concentrated mainly on vibration cracks and on zones and fractures produced by frost wedging.

Stratification

Stratification pertains to layering or bedding of geological materials that readily separate along bedding planes because of different kinds, arrangement, sizes, and assortment of material, or because of some geological processes — for example, interruption of deposition, or other changes affecting deposition.

Within layers, minor units about 5 mm or less in thickness are called *laminae*. A laminae-containing deposit is considered to be laminated.

Lamination of rock strata also constitutes a discontinuity in the geological medium. Hence, it is considered to be a mechanical weakness of rock *en masse*.

Separation is a break between beds along bedding planes.

As to rock engineering, horizontal, relatively thick rock strata are sometimes suitable for excavating small underground openings, such as driving small adits and narrow underground passageways, depending upon the intensity of the jointing. Here, in bridging the narrow span, the thick rock roof stratum acts like a beam or slab (Fig. 5-1a). Thin, fissured rock strata are unsuitable for construction of wide underground openings in rock (Fig. 5-1b). In such a case, the underground opening of a pointed arch roof design may be satisfactory (Fig. 5-1c).

Seams (Interbeds)

Stratified, horizontal rock layers on a thin seam of plastic clay, when undercut by an underground opening, have been observed to slide out of the vertical alignment into the opening (Fig. 5-2). This is so because of the removal of the lateral restraint brought about by the excavation of the rock.

Seams of plastic clay and of calcite may cause a stratified, dipped rock mass above the seam to slide down into the excavation (Fig. 5-3).

5-2. Fractures

A fracture is a fresh break in the continuity of a body of rock, not attended

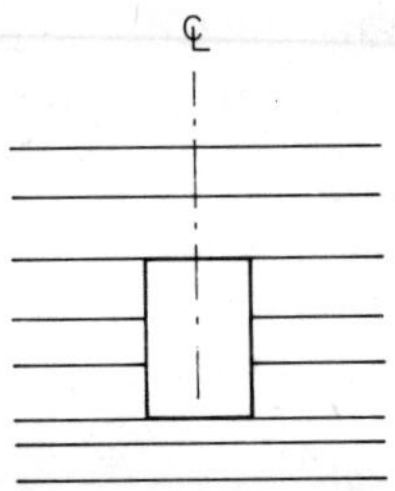

a) Bridging a narrow span of an underground opening by a thick competent rock roof stratum

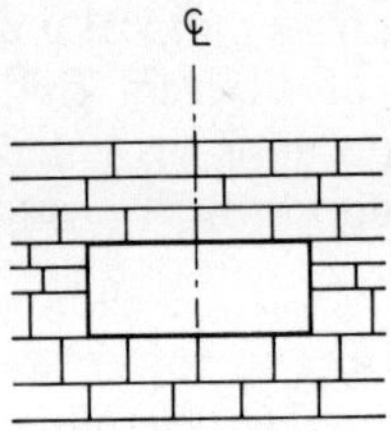

b) Thin incompetent jointed rock strata unsuitable for constructing a wide underground opening

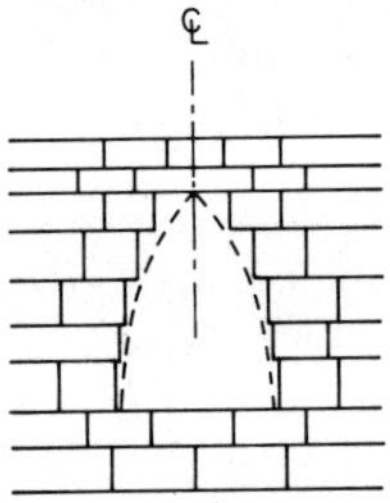

c) A pointed arch roof

Fig. 5-1
Underground openings in level rock strata

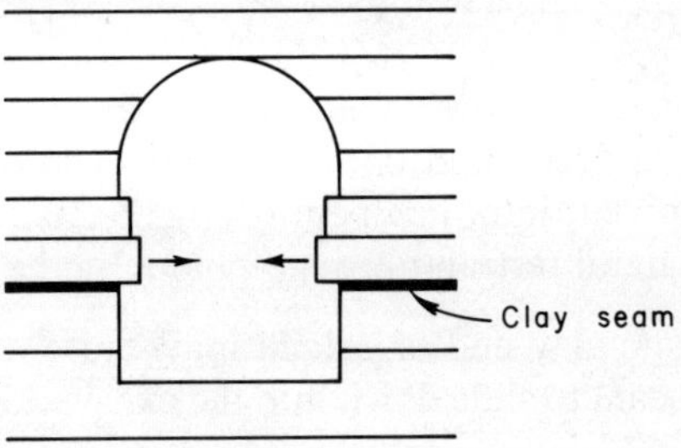

Fig. 5-2
Movement of rock strata over a clay seam

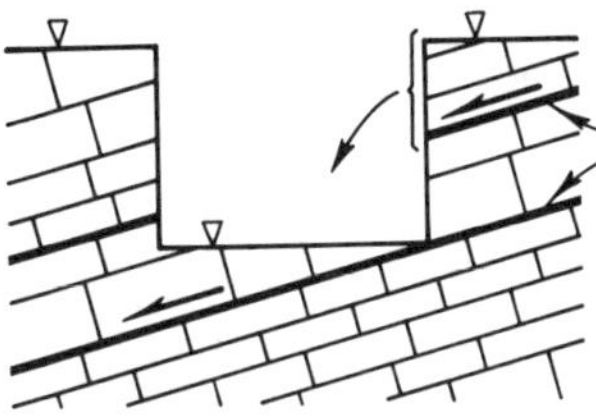

Fig. 5-3
Sliding over clay seams

by a displacement and not oriented in a regular system. Fractures may be open or closed but not bonded. Fractures are often man-made (for example, those caused by blasting).

Fractures may be kilometers in length, thousands of meters deep, and proportionally wide. Or else, fractures may be microscopic in all dimensions. Fractures in rock provide one of the easiest ways for water to enter.

The main types of fractures are *joints* and *faults*.

Generally, both joints and faults are surfaces of parting, or fractures—elements of rock weakness.

A *joint* is a crack or fracture in a rock along which no, or no noticeable, displacement has occurred. A *fault* is a fracture along which there has been some displacement.

Faults are named and classified according to inclination of the fault surface, and by the direction and relative movement of the rocks along the fault plane.

5-3. Joints

A joint is a simple fracture of geologic origin in the continuity of a rock mass. Joints in rock occur in a system of sets, but never singly.

If the walls of the joint are separated, the term fissure is sometimes used. Both joints and faults are geologic elements of practical interest to rock mechanics and rock engineering, i. e., foundation and hydraulic structures engineering.

Besides of ease or difficulty in working in rock; besides of *cost* of rock-works; and besides that groundwater flows along the rock and on the surfaces of joints and faults, rock alteration processes, too, take place. Much of the excessive leakage of water into an underground opening may take place along faults. This situation creates the problem of leaky and wet underground openings and tunnels.

Joints are formed by tensile stresses caused by contraction resulting from cooling of magma and of lava flows. The origin of planar joint sets is really not well known. However, almost every rock contains joints.

A joint can be *open* or *closed*. A *closed* joint is one whose walls are in contact. Closed joints may be almost invisible. Yet they constitue surfaces along which there is no resistance against separation. There are also tiny haircracks present in rocks.

Joints may be continuous or discontinuous.

Joints and joint systems have attracted the attention of builders ever since rock was used in engineering structures and as a structure-supporting material. In underground openings, joints affect the extent to which their sides, walls and the roof must be supported. Joints respond to explosions applied static and dynamic loads, and seepage of water. Because joints are one of the major causes of excessive *overbreak* of rock excavations (Fig. 5-4) and because joints constitute potential rock-slide hazards in unlined, undercut rock formations, and create water-trouble in rock excavations (Fig. 5-5), jointing always deserves careful attention and careful exploration.

Also, sand and other kinds of gouges tend to flow into an underground opening as the excavation is worked through a fault zone. A gouge is finely abraded material between the walls of a fault, the result of a grinding movement.

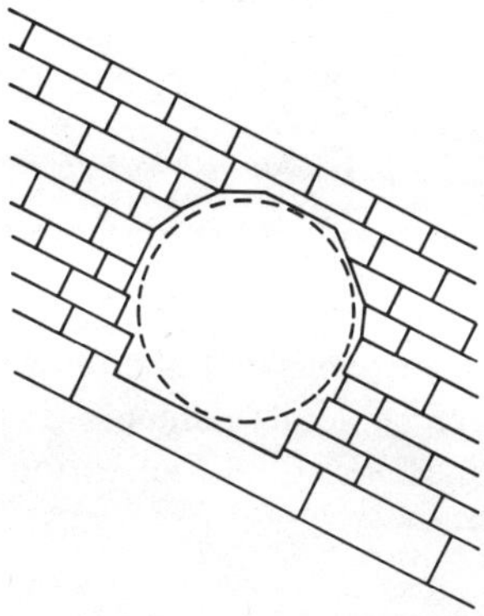

Fig. 5-4
Illustrating overbreak around a tunnel driven through a highly jointed stratified rock

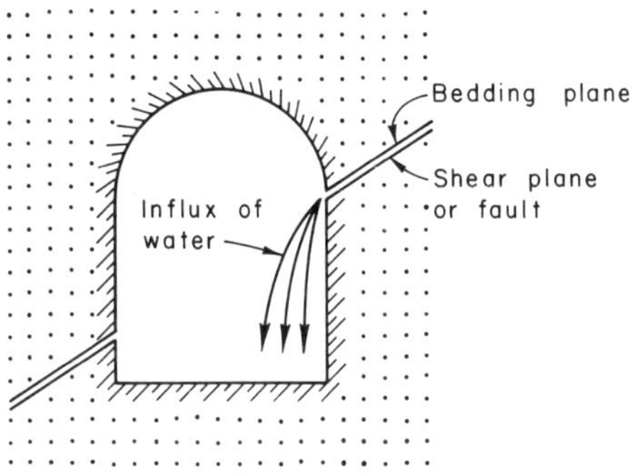

Fig. 5-5
Water influx from a bedding plane (or a joint, or fault zone) into an underground opening

Generally, joints subdivide rock masses into individual blocks to form a three-dimensional network that presents the rock as an uncemented, cohesionless aggregate of cuboid blocks somewhat comparable to closely fitted blocks in a dry masonry wall (imbrication).

The term *imbrication* means the lying of blocks lapped over each other in rock strata in a regular order so as to form an imbricate pattern surface. Some simple imbrication joint patterns are sketched in Fig. 5-6.

The influence of joints and faults on the failure of a rock mass is now widely recognized. Because the geometry of the joint pattern has a decisive effect on the strength of the rock mass to resist structural loads, the jointing patterns require some kind of characterization.

The need for rating and characterization of jointing and fissuring of rocks has been voiced at the First International Congress of Rock Mechanics held in Lisbon in 1966.

Assuming that in some instances the rock mass may be considered as being fractured in a regular network of rock blocks (which blocks are sometimes imbricated), Rocha (21) applied the concept of *imbrication i* for characterizing jointing in rocks (Fig. 5-6).

In particular, the more imbricated the blocks, the higher their shear strength, because rupture surfaces have not only to follow the joints, but also to cut the material of the blocks. A jointing system determines the ease of rock excavation by means of a power shovel or by means of explosives. The spacing of joints in quarries determines the largest size of blocks of sound rock that can be quarried for cut block stones for construction purposes.

Because joints are of geological origin, there are usually some weathering and decompositon products (gouge) on the joint surfaces which in some instances may bond the joint.

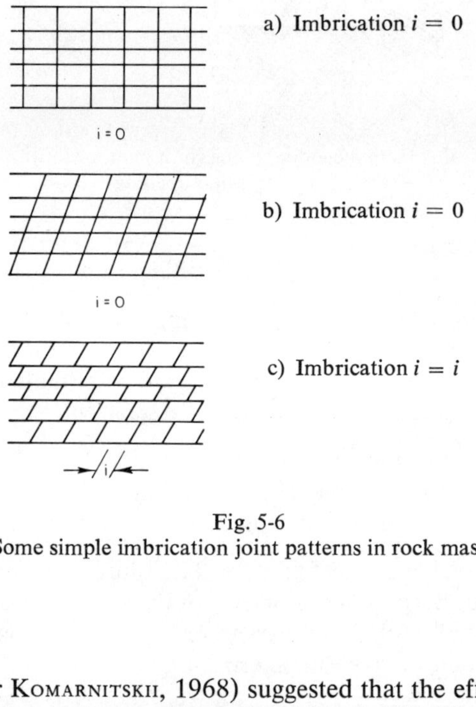

Fig. 5-6
Some simple imbrication joint patterns in rock masses

Popov (after Komarnitskii, 1968) suggested that the effect of jointing in rocks on their strength property may be characterized by a *coefficient of weakness*. It is the ratio of the cohesive force in the weak rock to the cohesive force in the same rock with no visible traces of such weakness. On the basis of experimental data, Popov devised a table (Table 5-1) showing the coefficient of weakness of rocks as a function of rock jointing characteristic.

5-4. Faults

If the rock masses on opposite sides of a rupture are relatively displaced or shifted, the break is called a fault. Thus, a fault is a fracture in a rock formation along which there has been movement parallel to the surface of the fracture. Faults occur in regions of tectonic disturbance. A fault, as a result of tectonic disturbance of a geological material, is a break or fracture

TABLE 5-1

Coefficient of Weakness of Rocks as a Function of Rock-Jointing Characteristics

(after S. I. Popov)*

Jointing characteristics in rock	Coefficient of Weakness	
	Limiting values	Average values
1	2	3
Dense network of fractures, in all directions of layered rock to individual uncemented blocks	0.0 −0.01	0.0005
Dense network of open fractures in all directions	0.001 − 0.02	0.005
Dense jointing	0.01 −0.04	0.02
Jointing above average	0.04 −0.08	0.06
Average jointing (open and closed fractures every 20–30 cm)	0.08 −0.12	0.1
Rocks with below-average jointing	0.12 −0.9	0.2
Network of deep joints every 30–50 cm; insignificant number of open fractures	0.3 −0.4	0.35
Little-jointed rocks; closed fractures	0.4 −0.6	0.5
Microfractures almost absent	0.6 −0.8	0.7
Monolithic rock with no sign of jointing	0.8 −1.0	0.9

* This table is reprinted with the permission of the PLENUM PUBLISHING CORPORATION from N. I. Komarnitskii, *Zones and Planes of Weakness in Rocks and Slope Stability*, p. 15, Table 1. Translated from the Russian by Consultants Bureau, New York, in 1968, and copyrighted by the Plenum Publishing Corporation.

(or zone of fractures or zone of weakness) in the continuity of a body or a system of rock (of a geological formation of rocks).

Mechanically, faults are shear failures of rock which result from tensile, compressive, or torsional stresses acting on a rock mass. Along the fracture or plane of failure, or fault plane, viz., shear plane, there takes place a relative shear displacement of the adjoining rock masses parallel to one another.

The shear displacement may be a few centimeters long, or run as long as many kilometers.

If the break is complex and the fractures occupy a wide strip, the fault plane becomes a fault zone.

In some instances the rock adjacent to faults is completely crushed. Such rock constitutes a *crushed zone*.

A fault sets in where the continuity of rock (or soil) layers is interrupted by failure in shear of the rock due to compressive forces, i. e., when the rock is strained past the breaking point and yields along a crack or series of cracks, so that corresponding points on the two sides become distinctly offset (Fig. 5-7a). One side of the so-sheared rock mass may rise or sink, or move laterally with respect to the other, depending on the nature of disturbance.

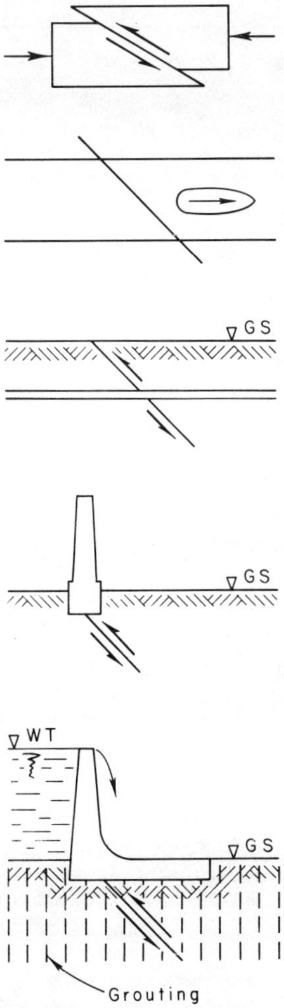

a) Failure in shear

b) Fault across a waterway

c) A utility line across a fault

d) A bridge pier over a fault

e) A dam built across a fault

Fig. 5-7
Structures across faults

Faults may be vertical, horizontal, or a combination thereof. Some fault structures are illustrated in Fig. 5-8.

In rock engineering, one distinguishes between *active or live faults* and *passive or dead faults*. An active fault is one in which movements or shear displacements have occured along it during the time of recorded history of man. In contradistinction, a passive fault is one in which no recorded slipping in the recorded history of man has taken place, but remains in static

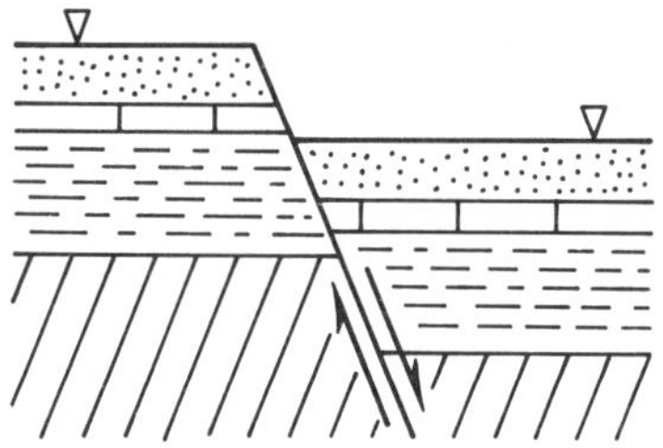

a) Normal fault. The hanging-wall block has moved relatively downward

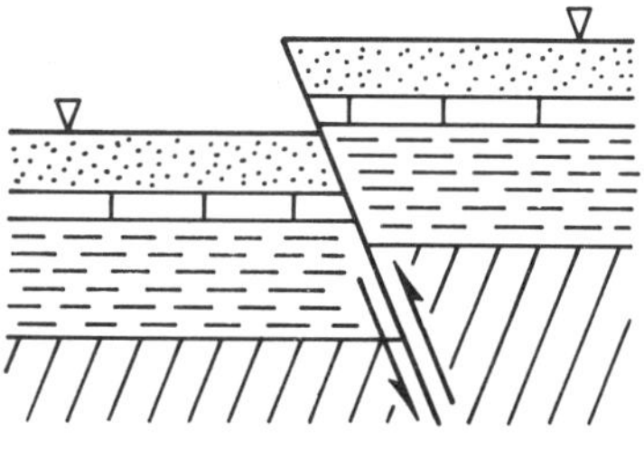

b) Reverse fault or thrust fault. The hanging-wall block has moved relatively upward

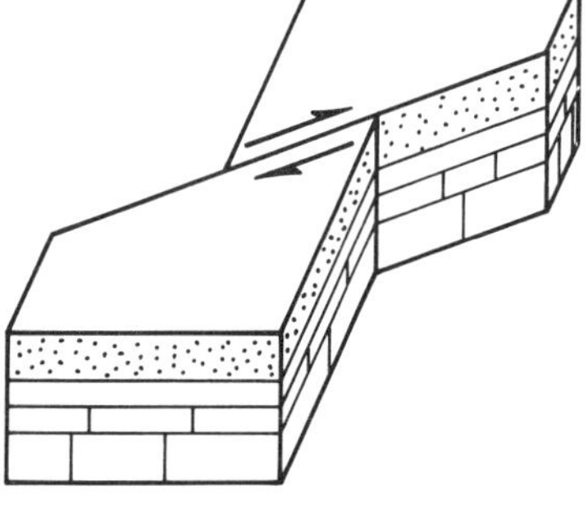

c) Horizontal displacement fault (strike-slip fault). Horizontal strata show no vertical displacement

Fig. 5-8
Principal kinds of faults

condition. The latter, however, is not a certainty because earthquake vibrations may cause the fault to open or to become active. Under structural load, static or dynamic, the rock mass can slide easily along the fault planes or surfaces.

In seismic regions, faults give rise to an additional problem; namely, when built across faults, structures such as tunnels, dams, bridges, and pipelines may experience, and may have experienced, distress and/or total destruction if subsequent movements of faults would occur, or have occured (Figs. 5-7b to 5-7e). Also, earthquake tremors and man-induced vibrations may cause a fault to open or to become active. Hence, in geotechnical engineering, faults constitute an undesirable hazard in working in rock.

5-5. Folds

A layered rock system, originally in a horizontal position, may be folded into folds of successive degrees of intensity of the partaking forces. Folds result from forces acting tangentially to the earth's surface (Fig. 5-9). The nature of the rock and its strength determine whether the strata of a rock formation will rupture or fold. The folds may assume a wide variety of shapes. An upwarped or upfolded segment of the earth's crust is an anticline. A downwarped or downfolded segment is a syncline.

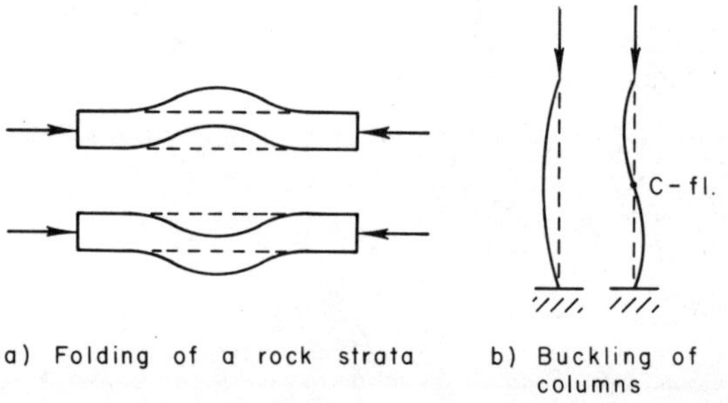

Fig. 5-9
Formation of folds

In geotechnical engineering, the synclines may cause problems of water accumulation. Also, rock folding is usually accompanied by fissures in anticlines and synclines. Along the crest of anticlines tension cracks are usually formed in the upper fibers where the outer rock strata are under tension (Fig. 5-10). In synclines, the lower fibers of the rock strata are under tension, hence the cracks at the bottom (Fig. 5-11).

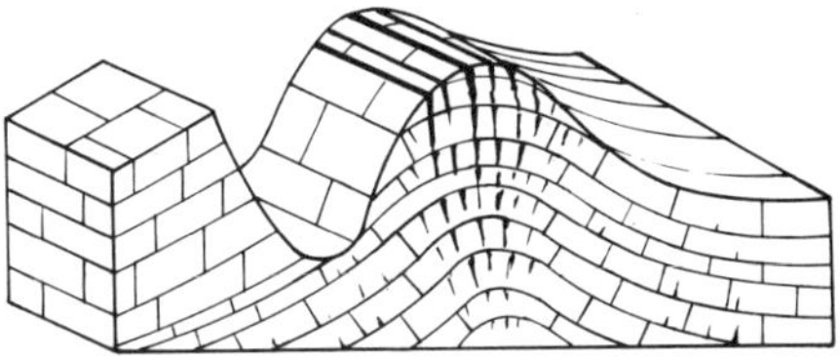

Fig. 5-10
Cracks in an anticline

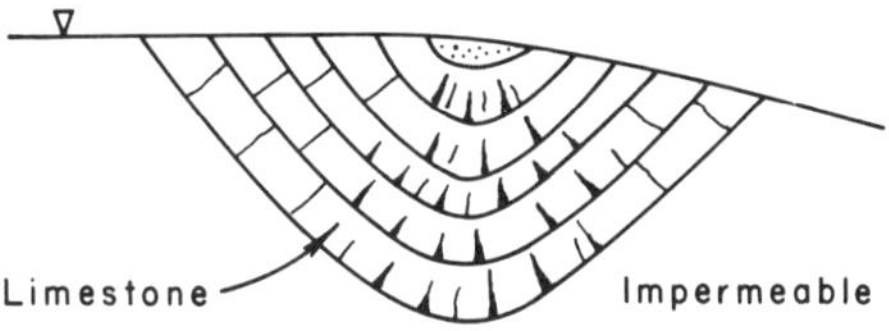

Fig. 5-11
Cracks in a syncline

Sliding of rock masses, too, is associated with folds.

Besides mechanical defects in folded rocks, one should also consider the effect of surcharge pressure imparted on a structure by folded rocks. This may be vizualized from the following examples.

If the alignment of a tunnel is traced normal to the strike under an anticline or syncline of a folded rock system, there is a variation in rock surcharge pressure p on the tunnel from rock masses from above. These variations are qualitatively sketched in Figs. 5-12 and 5-13. Figure 5-12 shows variation in surcharge pressure on a tunnel driven through an anticline. Notice the arching effect of the convexly curved rock stratification on the magnitude of pressure at the ends (basis) of the arch at both ends and at the middle of the tunnel.

If a tunnel is driven through a syncline (Fig. 5-13), large rock pressures are imparted in the middle part of the tunnel. Here the large stresses in the rock strata transmitted to the tunnel may complicate the driving and construction of the tunnel technically as well as economically. Also, large quantities of water may accumulate in the syncline. Hence, saturated layers can be anticipated here, requiring elaborate drainage systems.

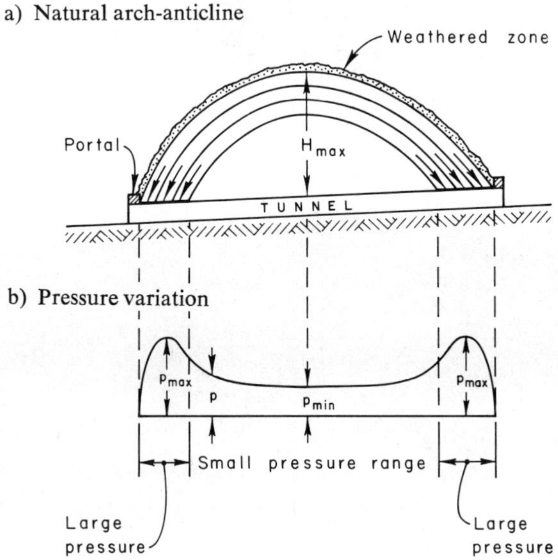

Fig. 5-12
Variation in surcharge pressure on a tunnel driven through an anticline

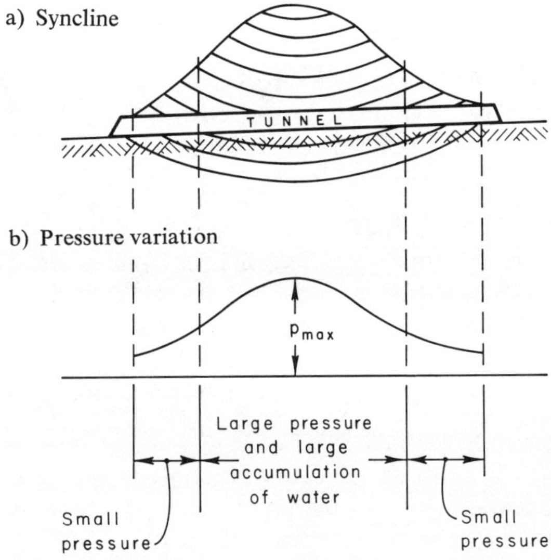

Fig. 5-13
Variation in surcharge pressure on a tunnel driven through a syncline

Figure 5-14 shows possible positions of a tunnel in horizontal rock formations.

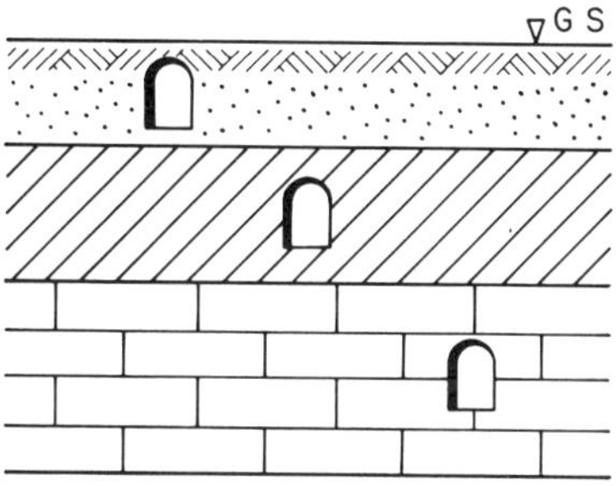

Fig. 5-14
Tunnel positions in horizontal strata

Figure 5-15 shows positions of a tunnel parallel with the strike in steeply dipping strata.

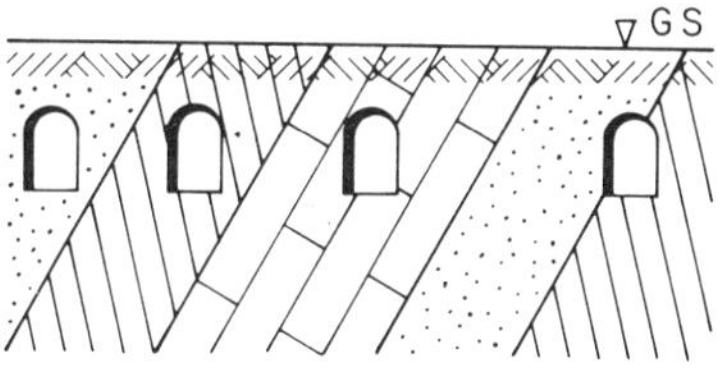

Fig. 5-15
Tunnel positions in steeply dipping strata

Figure 5-16 illustrates three possible variants of tunnel location:

1. In the syncline; large pressures and large accumulation of water; saturated rock strata may be anticipated;
2. In the anticline; natural arch; least pressure on tunnel; and
3. Increased stresses in layers, which stresses complicate the driving and construction of the tunnel.

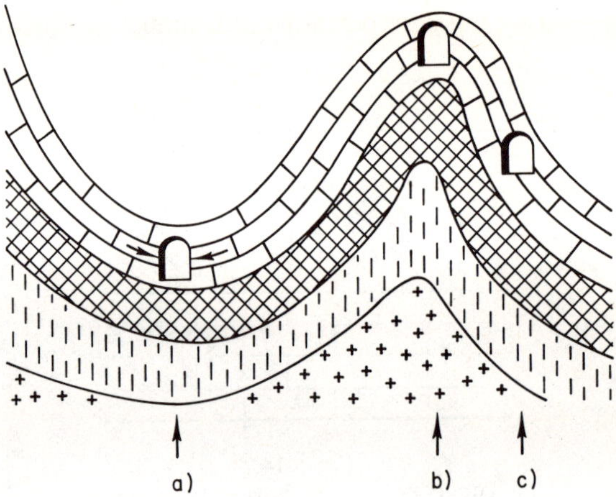

Fig. 5-16
Locations of tunnels in syncline and anticline

The conditions of folded rock systems and their relation to stability of the water-impounding massive dams are sketched in Fig. 5-17. In the case of a dam founded on an anticline (Fig. 5-17a) one expects a relatively small uplift pressure. However, the dam appears less safe against lateral sliding. The case where a dam is founded on a syncline (Fig. 5-17b) presents an unfavorable uplift condition for the dam; it is, though, relatively safe against sliding.

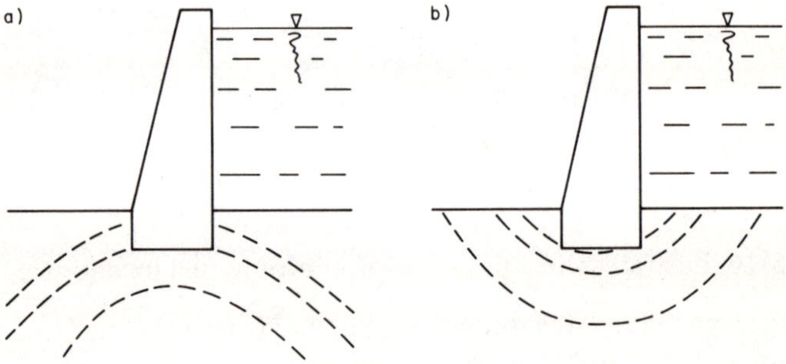

Fig. 5-17
Effect of rock folding on performance of a dam

Cavities and Voids

Sinkholes and other sorts of cavities are most commonly formed in limestone, gypsum, and salt. The collapsing of roofs of the cavities by structural loads means distress to the structure if founded on such a cavity-containing rock. Thus, cavities consitute mechanical defects of the rocks concerned.

Also, large voids and great porosity of rocks mean weaknesses requiring grouting or chemical injection, especially for hydraulic structures such as hydraulic power plants, aqueous pressure tunnels, and the like.

Porosity

Pores or voids in rock, too, are very important forms of non-uniformity of a rock structure. The largest porosities have been observed in sedimentary rocks. Porosity decreases the strength of the rock because, at the places of pores or voids there is a lack of bond (cementation) in the rock.

The foregoing review of mechanical weaknesses of rocks reveals that rock properties, like soil properties, vary from place to place and from stratum to stratum. Also, they vary from sample to sample even if the sample is obtained from the same rock formation. In this respect, and considering the many kinds of weaknesses a rock material may contain, one discovers that fissured, jointed, faulted, and folded rocks as a construction material for transmitting and distributing of structural loads to the rock ground, especially where the transmission of the horizontal components of inclined loads are concerned, are a nonhomogeneous, anisotropic material of bewildering complexity.

Hence, the nature of the rock mass and its complex structure call increasingly for specialized approaches in rock mechanics and rock engineering for design of engineering structures on and in rock.

5-6. Summary on Rock Mass Weaknesses

The significance of rock in dam and foundation engineering comes to the fore from the fact that more than 80 % of failures of massive dams are attributable to insufficient or incompetent strength of the rock, and/or deficient laying of foundations of such structures.

The rock should be accepted by the engineer as given by Mother Nature in its natural state with all of its own weaknesses, as determined by stratification, cracks, fissures and other mechanical rock defects, weathering, and the like. Of course, if economically and technically feasible, the strength properties of the rock can be improved, thus increasing the stability of the dam or foundation or underground opening, whichever the case may be.

Foremostly, it is the macroscopic structural anisotropy of the rocks, as governed by stratification, fissuring, and jointing, which frequently causes considerable deformations of the underground rock, but for which the analytical comprehension is not satisfactorily possible as yet.

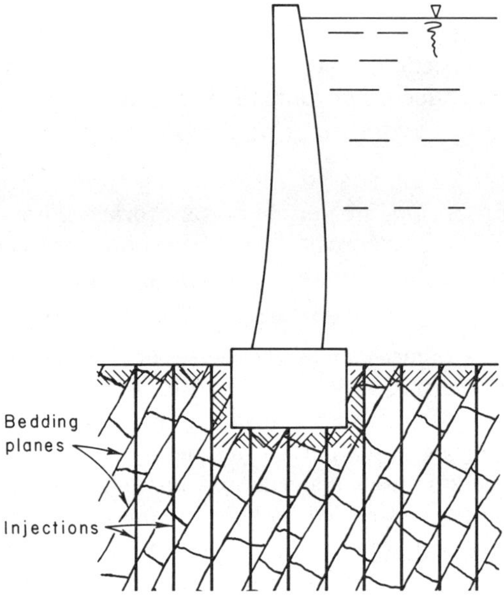

Fig. 5-18
Rock-dam system

Based on what was said above, let us fix a mental picture of a water-impounding concrete dam founded on or in a rock. A dam (barrage) and its supporting rock (Fig. 5-18) should form a static, integral unit; i. e., a barrage should be united with its supporting rock by a "solid marriage." Unfortunately, however, it usually turns out to be an unequal marriage, because:

She, whose name is Madame Gracia Archy Barrage, is young, formfit, graceful, and beautiful in shape.

He, whose name is Señor Rock, is usually age-old and weathered.

She: Smooth, slender, well-built, and of good-looking stature.

He: His interior is disturbed by tectonic processes in the past; his face is creased and deeply furrowed by chasms, caverns, cracks, and fissure wrinkles, or swollen up with folds.

She: Is getting quickly tall and strong (grown-up) and lives frequently on a large pedestal; and according to the opinions of geotechnical engineers and concrete technologists, the finest for her is just good enough.

He: Upon him rests all the applied loads. He bears them patiently, and fortunately for him, he is smarter than all the calculations and analyses, and he knows how to help himself (visualize stress distribution in rock). However, if the rock is weak, it may be given antiweakness medicine-injections. If in spite of this "medical help" the Rock constitution loses his internal balance (equilibrium), the rock simply breaks, ruptures — he is finished. Subsequently, and tragically, she also, Madame Barrage, loses her existence.

Nowadays, earth and concrete dams become higher, longer, bigger, thus increasing in their weight. And so increase, also, the forces which are transmitted into the underground. Over and above this, engineers are faced with the necessity to select and adopt more unfavorable construction sites because most of the best sites have been built up already. Therefore, an analytical treatment of the stresses in the underground rock medium, a technological stress measurement in situ, and testing of rock strength become now, more and more an urgent necessity than before.

Relative to the dam, underground opening, and structural foundation-supporting rock, one reckons with the following forces:

1. Self-weight of structure,
2. Weight of rock,
3. Geological forces brought about by tectonic processes and seismic action,
4. Hydrostatic uplift,
5. Hydrostatic shear, induced by the load of the impounded water,
6. Seismic forces, and
7. Reactions brought about by the structure.

The higher the dam, the greater are these forces .The highest concrete dam in the world today is the Grande Dixence dam, 281 m high (in Switzerland).

Influence of Rock on Stability of Dams and Foundations

The basic concept in rock mechanics studies and design pertaining to subsurface engineering in rock (domes, tunnels, shafts, foundations) is the consideration of the rock and the structure as a functionally and statically coherent, integral entity. The prerequisite for this, besides knowledge of geological structure, stratigraphy, and rock weaknesses, is knowledge of the physical and mechanical properties of the rock under consideration.

The mutual interaction of rock and structural foundation systems is of utmost importance and should not be underestimated. The mutual interaction forms the basis for giving the form to the foundation and to its proportioning, viz., sizing the structure itself. This basis, in its turn, depends upon the strength properties of the foundation material itself, and the foundation-supporting rock. Also, in tunnel and shaft engineering, the rock pressure on such underground-opening structures, and its stress and strain influence on the rock in the vicinity of the tunnel and/or shaft opening, come to the fore. Thus, rock mechanics is a discipline subject to all concepts of rock testing, technical evaluation, and engineering application. The concepts are based on exact bases of theoretical and applied mechanics, soil mechanics principles, and foundation engineering.

REFERENCES

1. Andreae, C., 1953, "Die Temperaturprognose im Tunnelbau," *Schweizerische Bauzeitung,* pp. 41–42.
2. ASCE – *Symposium on Underground Rock Chambers,* an ASCE National Meeting held January 13–14, 1971, at Phoenix, Arizona. New York, N. Y.: American Society of Civil Engineers.
3. ASME – *Rock Mechanics Symposium,* A. L. Sikarskie (editor), 1973, The Winter Annual Meeting of the ASME, held November 11–15, 1973, at Detroit, Mich., The American Society of Mechanical Engineers.
4. ASTM – Special Technical Publication 420, 1966, *Testing Techniques for Rock Mechanics.* Philadelphia, Pa.: American Society for Testing and Materials.
5. ASTM – Special Technical Publication 429, 1967, *Determination of Stress in Rock,* A State-of-the-Art Report. Philadelphia, Pa.: American Society for Testing and Materials.
6. ASTM – Special Technical Publication 477, 1969, *Determination of the In-Situ Modulus of Deformation of Rock.* Philadelphia, Pa.: American Society for Testing and Materials.
7. ASTM – Special Technical Publication 479, 1970 (5th ed.), *Special Procedures for Testing Soil and Rock for Engineering Purposes.* Philadelphia, Pa.: American Society for Testing and Materials.
8. ASTM – Special Technical Publication 483, 1971, *Sampling of Soil and Rock.* Philadelphia, Pa.: American Society for Testing and Materials.
9. ASTM – Special Technical Publication 554, 1974, *Field Testing and Instrumentation of Rock.* Philadelphia, Pa.: American Society for Testing and Materials.
10. ASTM – Standard Definitions of Terms and Symbols Relating to Soil and Rock Mechanics, D653–67 (Reapproved 1973), *1977 Annual Book of ASTM Standards,* Part 19. Philadelphia, Pa.: American Society for Testing and Materials, 1977.
11. ASTM – Standard Test Method for Modulus of Rupture of Natural Building Stone, Designation C99–52 (Reapproved 1970), *1977 Annual Book of ASTM Standards,* Part 19. Philadelphia, Pa.: American Society for Testing and Materials, 1977, pp. 4–6.

12. ASTM – Standard Test Method for Compressive Strength of Natural Building Stone, Designation C 170–50 (Reapproved 1970). *1977 Annual Book of ASTM Standards,* Part 19. Philadelphia, Pa.: American Society for Testing and Materials, 1977, pp. 16–18.

13. ASTM – Standard Test Method for Unconfined Compressive Strength of Intact Rock Core Specimens, American National Standard ANSI/ASTM D2938–71a, *1977 Annual Book of ASTM Standards,* Part 19. Philadelphia, Pa.: American Society for Testing and Materials, 1977, pp. 374–375.

14. ASTM – Standard Test Method for Direct Tensile Strength of Intact Rock Core Specimens, ANSI/ASTM D2936–71, *1977 Annual Book of ASTM Standards,* Part 19. Philadelphia, Pa.: American Society for Testing and Materials, 1977, pp. 366–368.

15. ASTM – Standard Test Method for Triaxial Compressive Strength of Undrained Rock Core Specimens without Pore Pressure Measurements, ASTM Designation D2664–67 (Reapproved 1974), *1977 Annual Book of ASTM Standards,* Part 19. Philadelphia, Pa.: American Society for Testing and Materials, 1977, pp. 327–330.

16. Baidyuk, B. V., 1967, *Mechanical Properties of Rocks at High Temperatures and Pressures.* New York, N. Y.: Consultants Bureau.

17. Barney, K. R., 1960, "Madison Canyon Slide," *Civil Engineering,* August 1960, pp. 72–75.

18. Birch, F., 1972, "Compressibility, Elastic Constants," in *American Institute of Physics Handbook,* 3rd ed. New York, N. Y.: McGraw-Hill Book Company, Table 7–15 on Rocks: Elastic Constants, pp. 100–170.

19. Boussinesq, J. V., 1885, *Application des potentiels à l'étude de l'equilibre et du mouvement des solides élastiques.* Paris: Gautier-Villars.

20. Brace, W. F., 1971, "Microcracks in Rock Systems," *Proceedings Southampton 1969 Civil Engineering Materials Conference.* London: John Wiley and Sons, Inc., Part I, pp. 187–204.

21. Brekke, T. L., and F. A. Jörstad (editors), 1970, *International Symposium on Large Underground Openings, Oslo, 1969.* Oslo: Universitetsforlaget.

22. U.S. Bureau of Reclamation, 1954, *Physical Properties of Typical Foundation Rocks,* Concrete Laboratory Report SP39.

23. Casagrande, A., and R. E. Fadum, 1939, *Notes on Soil Testing for Engineering Purposes.* Cambridge, Mass.: Harvard University Publication No. 268, 1939/40, Fig. II, p. 23.

24. Clark, S. P. Jr., 1966, *Handbook of Physical Constants,* The Geological Society of America, Memoir 97.

25. Coulomb, C. A., 1776, "Essai sur une application des régles de maximis et minimis à quelques problémes de statique rélatifs à l'architecture." Paris: De l'Imprimerie Royale: *Mémoires de la mathématique et de physique,* présentés à l'Académie Royale des Sciences, par divers Savants, et lûs dans ses Assemblées, Vol. 7, Année 1773, pp. 346–384.

26. Duckworth, W. H., 1951, "Precise Tensile Properties of Ceramic Bodies," *Journal of the American Ceramics Society,* Vol. 34, No. 1, pp. 1–9.

27. Eshbach, O. W., (editor), 1952, *Handbook of Engineering Fundamentals.* New York, N. Y.: John Wiley and Sons, Inc., pp. 13–23 to 13–24.

28. Everell, M. D., Herget, G., Sage, R., and D. F. Coates, 1974, "Mechanical Properties of Rocks and Rock Masses," 3rd International Congress on Rock Mechanics at Denver, Colo., U.S.A.

29. Fairhurst, C., 1961, "Laboratory Measurement of Some Physical Properties of Rock," *Proceedings 4th Symposium on Rock Mechanics,* held 1961 at University Park, Pa., pp. 105–118.

30. FAIRHURST, C., 1964, "On the Validity of the 'Brazilian' Test for Brittle Materials," *International Journal for Rock Mechanics and Mining Science,* Vol. I, No. 4, pp. 535–546.

30a. FAIRHURST, C., 1964, "Measurement of In-Situ Rock Stresses, with Particular Reference to Hydraulic Fracturing," *Rock Mechanics and Engineering Geology,* Vol. 2, No. 3/4, pp. 129–147.

31. FAIRHURST, C., 1976, "In Situ Stresses Existence, Determination and Data Interpretation," *ASCE Rock Excavation Seminar,* held October 1976 in New York, N. Y., ASCE Metropolitan Section, Foundations and Soil Mechanics Group.

32. FARMER, I. W., 1968, *Engineering Properties of Rocks.* London: E. & F. N. Spon Ltd.

33. GOODMAN, R. E., VAN, K., and F. E. HEUZE, 1972, "Measurement of Rock Deformability in Boreholes," in *Basic and Applied Rock Mechanics,* Proceedings of the 10th Symposium on Rock Mechanics, New York: Intersociety Committee for Rock Mechanics, The Society of Mining Engineers of the American Institute of Mining Engineers in New York, pp. 523–555.

34. GRIFFITH, J. H., 1936, *Thermal Expansion of Typical American Rocks.* Ames, Iowa: Iowa State College of Agriculture and Mechanic Arts, Iowa Engineering Experiment Station, Vol. 35, No. 19, October 7, 1936, pp. 19–20.

35. GRIFFITH, J. H., 1937, *Physical Properties of Typical American Rocks.* Ames, Iowa: Iowa State College of Agriculture and Mechanic Arts, Iowa Engineering Experiment Station, Bulletin 131, Vol. 35, No. 42.

36. GRIFFITH, A. A., 1921, "The Phenomena of Rupture and Flow in Solids," *Philosophical Transactions,* Vol. 221, pp. 163–198.

36a. GRIFFITH, A. A., 1924, "Theory of Rupture," Proceedings First International Congress on Applied Mechanics. Delft: J. Waltham, Jr. Press, pp. 55–64.

37. HAIMSON, B. C., 1974, "A Simple Method for Estimating In Situ Stresses at Great Depth," in *Field Testing and Instrumentation of Rock.* Philadelphia, Pa.: ASTM Special Technical Publication 554, pp. 156–182.

38. HARD, H. C. et al., 1972, *Flow and Fracture of Rocks: The Griggs Volume.* Washington, D. C.: American Geophysical Union, Geophysical Monograph 16.

39. HEARD, H. C., 1963, "The Effect of Large Changes in Strain Rate in the Experimental Deformation of Rocks," *Journal of Geology,* Vol. 71, p. 162.

40. HIX, C. F. Jr., and R. P. ALLEY, 1958, *Physical Laws and Effects.* New York, N. Y.: John Wiley and Sons, Inc., p. 64.

41. HOBBS, D. W., 1964, "The Tensile Strength of Rocks," *International Journal for Rock Mechanics and Mining Science,* Vol. I, pp. 385–396.

42. HORVATH, J., 1965, "A New Approach to the Determination of Stresses in the Earth's Crust and Strata Pressure on Tunnel Linings," *International Journal for Rock Mechanics and Mining Science,* Vol. 2, pp. 327–340.

43. International Society for Rock Mechanics, Commission on Standardization of Laboratory and Field Tests, 1972, *Suggested Methods for Determining the Uniaxial Compressive Strength of Rock Materials and the Point Load Strength Index.* Lisbon: Committee on Laboratory Tests, Documents No. 1 and No. 2, October and November, 1972.

44. International Society for Rock Mechanics, 1972, *Proceedings Percolation through Fissured Rock,* a Symposium held in 1972 at Stuttgart. Essen: Deutsche Gesellschaft für Erd- und Grundbau.

45. JAEGER, C., 1963, "The Malpasset Report," *Water Power,* February 1963, pp. 55–61.

46. JAEGER, J. C., and E. R. HOSKINS, 1966, "Rock Failure under the Confined Brazilian Test," *Journal of Geophysical Research,* Vol. 71, No. 10, pp. 2651–2659.

47. JAEGER, J. C., and N. G. COOK, 1969, *Fundamentals of Rock Mechanics*. London: Chapman and Hall, p. 82.

48. JAEGER, J. C., 1976, *Fundamentals of Rock Mechanics* (2nd ed.). London: Chapman and Hall; New York, N. Y.: John Wiley and Sons, Inc.

49. JOHNSON, A. M., 1970, *Physical Processes in Geology*. San Francisco, Calif.: Freeman, Cooper and Company, p. 202.

50. JUMIKIS, A. R., 1956, "Rupture Surface in Sand under Oblique Loads," *Proceedings ASCE,* Soil Mechanics and Foundations Division, Paper 861, Vol. 82, pp. SM-1 to SM-6.

51. JUMIKIS, A. R., 1965, *Stability Analyses of Soil-Foundation Systems*. New Brunswick, N. J.: Rutgers University, College of Engineering, Bureau of Engineering Research, Engineering Research Publication 44 (55 pages).

52. JUMIKIS, A. R., 1967, *Introduction to Soil Mechanics*. NewYork, N. Y.: D. Van Nostrand Company, pp. 341–344.

53. JUMIKIS, A. R., 1969, *Experimental Studies on Moisture Transfer in a Silty Soil upon Freezing as a Function of Porosity*. New Brunswick, N. J.: Rutgers University, College of Engineering, Engineering Research Bulletin 49, pp. 30–32, 40.

54. JUMIKIS, A. R., 1973, *Settlement Tables for Centrically Loaded Rigid Circular Footings on Multilayered Soil Systems*. New Brunswick, N. J.: Rutgers University, College of Engineering, Engineering Research Bulletin 54, p. 21.

55. JUMIKIS, A. R., and A. A. JUMIKIS, 1975,, *Red Brunswick Shale and its Engineering Aspects*. New Brunswick, N. J.: Rutgers University,College of Engineering, Engineering Research Bulletin 55, p. 51.

56. JUMIKIS, A. R., and W. A. SLUSARCHUK, 1973, "Electrical Parameters of Some Frost-Prone Soils," in *Advanced Concepts and Techniques in the Study of Snow and Ice Resources,* a United States contribution to the International Hydrological Decade. Washington, D. C.: National Academy of Sciences, 1974, pp. 765–781.

57. JUMIKIS, A. R., 1977, *Thermal Geotechnics*. New Brunswick, N. J.: Rutgers University Press, pp. 320–238.

58. KÁRMAN, VON, T., 1911, "Festigkeitsversuche unter allseitigem Druck," *Zeitschrift des Vereins Deutscher Ingenieure,* Vol. 55, pp. 1749–1757.

59. KIERSCH, G. A., 1964, "Vaiont Reservoir Disaster," *Civil Engineering,* Vol. 34, pp. 32–39.

60. KRYNINE, D. P., and W. R. JUDD, 1957, *Principles of Engineering Geology and Geotechnics*. New York, N. Y.: McGraw-Hill Book Company, Inc., pp. 23–24.

61. LEEMAN, E. R., and D. J. HAYES, 1966, "A Technique for Determining the Complete State of Stress in Rock Using a Single Borehole," *Proceedings First Congress of the International Society of Rock Mechanics,* held September 25 to October 1, 1966, at Lisbon, Vol. 2, Paper 4.3, pp. 17–24.

62. MANGHNANI, M. H., SCHREIBER, E., and N. SOGA, 1968, "Use of Ultrasonic Interferometry Technique for Studying Elastic Properties of Rocks," *Journal of Geophysical Research,* Vol. 73, No. 2, pp. 824–826.

63. MENCL, V., 1966, "Mechanics of Landslides with Noncircular Slip Surfaces with Special Reference on the Vaiont Slide," *Géotechnique,* Vol. 16, No. 4, pp. 329–337.

64. MOHR, O., 1900, in *Zeitschrift des Vereins Deutscher Ingenieure,* Vol. 44.

65. MOOS, A., and F. DE QUERVAIN, 1948, *Technische Gesteinskunde*. Basel: Birkhäuser.

66. MÜLLER, L., 1964, "The Rock Slide in the Vaiont Valley," *Felsmechanik und Ingenieurgeologie,* Vol. 2, pp. 148–228.

67. MÜLLER, L., 1968, "New Considerations on the Vaiont Slide," *Rock Mechanics and Engineering Geology,* Vol. 6. Wien–New York, N. Y.: Springer-Verlag.

68. MÜLLER, L., 1965, "Safety in Rock Engineering" ("Die Sicherheit im Felsbau"), 15th Colloquium of the Austrian Regional Group of the International Society for Rock Mechanics, held September 24–25, 1964, at Salzburg. Wien: Springer.

69. MÜLLER, L., (editor), 1974, *Rock Mechanics,* A course held at the Department of Mechanics of Solids of the International Center for Mechanical Sciences at Udine, Italy. Wien–New York, N. Y.: Springer.

70. NÁDAI, A., 1950, *Theory of Flow and Fracture of Solids,* Vol. 1 (2nd ed.). New York, N. Y.: McGraw-Hill Book Company, Inc., Chapter 24, p. 379.

71. OBERT, L., WINDES, S. L., and W. I. DUVALL, 1946, *Standardized Tests for Determining the Physical Properties of Mine Rock,* Bureau of Mines Report 3891.

72. OBERT, L., 1962, "In-Situ Determination of Stress in Rock," *Mining Engineering,* Vol. 14.

73. OBERT, L., and W. I. DUVALL, 1967, *Rock Mechanics and the Design of Structures in Rock.* New York, N. Y.: John Wiley and Sons, Inc.

74. PATTON, F. D., 1968, Lecture notes on "The Determination of Shear Strength of Rock Masses," Lecture given on May 7, 1968, at the Seminar on Rock Mechanics in Civil Engineering Practice to the Metropolitan Section of ASCE, Foundations and Soil Mechanics Group, in New York City, p. 12.

75. PRICE, N. J., 1966, *Fault and Joint Development in Brittle and Semibrittle Solids.* Oxford: Pergamon Press.

76. *Proceedings Geotechnical Conference Oslo, 1967* on Shear Strength Properties of Natural Soils and Rocks. Oslo: Norwegian Geotechnical Institute, Vol. 1, 1967.

77. ROBERTSON, E. C., 1959, "Experimental Study of the Shear Strength of Rocks," *Bulletin of the Geological Society of America,* Vol. 66, October 1959.

78. ROCHA, M., 1966, "Determination of the Deformability of Rock Masses along Boreholes," *Proceedings First Congress of the International Society of Rock Mechanics,* held from September 25 to October 1, 1966, at Lisbon, Vol. 1, Paper 3.77, pp. 697–704.

78a. ROCHA, M. et al., 1966, "A New Technique for Applying the Method of the Flat Jack in the Determination of Stresses inside Rock Masses," *Proceedings First Congress of the International Society of Rock Mechanics,* held September 25 to October 1, 1966, at Lisbon, Vol. 2, Paper 4.10, pp. 57–65.

79. ROWE, P. W., 1964, "Importance of Free Ends in Triaxial Testing," *Proceedings ASCE,* Soil Mechanics and Foundations Division, Vol. 90, pp. 1–77.

80. SERAFIM, J. L., 1962, *Internal Stresses in Galleries.* Lisbon: Laboratorio Nacional de Engenharia Civil, Technical Paper No. 204.

81. SINGER, F. L., 1951, *Strength of Materials.* New York, N. Y.: Harper & Brothers, Publishers, p. 69.

82. SKEMPTON, A. W., 1966, "Bedding Plane Slip, Residual Strength and the Vaiont Landslide," *Géotechnique,* Vol. 16, No. 1, pp. 82–84.

83. *Smithsonian Physical Tables,* 1920, 7th ed., Washington, D. C., Vol. 71, No. 1, Publication 2359, p. 229.

84. STAGG, K. G., 1968, "In Situ Tests on the Rock Mass," in *Rock Mechanics in Engineering Practice,* K. G. STAGG and O. C. ZIENKIEWICZ (editors). London–New York, N. Y.: John Wiley and Sons, Inc., pp. 135–137.

85. Stanford Research Institute, 1966, "On the Strength of Rock," *SRI Journal,* No. 9, March 1966, pp. 12–14.

86. SZÉCHY, K., 1966, *The Art of Tunnelling*. Budapest: Akadémiai Kiadó, p. 76.

87. TERZAGHI, K., 1962, "Measurement of Stresses in Rock," *Géotechnique* (London), Vol. 12, No. 2, pp. 105–124.

88. THOMAS, L. J., 1973, *An Introduction to Mining: Exploration, Feasibility, Extraction, Rock Mechanics*. Sydney: Hicks, Smith and Sons.

89. TIMOSHENKO, S., and J. N. GOODIER, 1951, *Theory of Elasticity*, 2nd. New York, N. Y.: McGraw-Hill Book Company, Inc.

90. TRESCA, H., 1864, 1867, "Mémoire sur l'écoulement des corps solides soumis à des fortes pressions," Paris: *Comptes Rendus*, Académie des Sciences, July–December, Vol. 59, 1864, pp. 754–758; Vol. 64, 1867, p. 809.

91. VUTUKURI, V. S., LAMA, R. D., and S. S. SALUJA, 1974, *Handbook on Mechanical Properties of Rocks*, Vol. 1. Clausthal, Germany: Trans Tech Publications.

92. VUTUKURI, V. S., and R. D. LAMA, 1978, *Handbook on Mechanical Properties of Rocks*, Vol. II, Vol. III and Vol. IV. Clausthal, Germany: Trans Tech Publications.

OTHER RELATED REFERENCES

ASTM – Special Technical Publication 601, 1976, *National Symposium on Fracture Mechanics*, 9th Symposium held at the University of Pittsburgh, Pa., in 1975. Philadelphia, Pa.: Ameriacn Society for Testing and Materials.

BROCK, D., 1974, *Elementary Engineering Fracture Mechanics*. Leyden: Nordhoff International Publishing.

CORDING, E. J., 1968, "Stability of Large Underground Openings at the Nevada Test Site," Preprint of paper presented at the Second Space-Age Conference of the ASCE, Los Angeles, April 25, 1968.

DONATH, F. A., 1964, *Effect of Loading Rate on the Deformation Behavior of Rocks Subjected to Triaxial Compression*, ARO-D Report No. 3055: 1, July 31, 1964. New York, N. Y.: Columbia University (81 pages).

FAIRHURST, C., (editor), 1967, *Failure and Breakage of Rock, Proceedings 8th Symposium on Rock Mechanics*, American Institute of Mining Engineers, pp. 237–302.

FUMAGALLY, E., 1973, *Statistical and Geomechanical Models*. New York, N. Y.: Springer-Verlag.

GRETENER, P. E., 1965, "Can the State of Stress Be Determined from Hydraulic Fracturing Data," *Journal of Geophysical Research*, Vol. 70, pp. 6205–6215.

GRIGGS, D., 1939, "Creep of Rocks," *Journal of Geology*, Vol. 47, No. 3, pp. 225–251.

GRIGGS, D., and J. HANDIN, 1960, "Observations on Fracture and a Hypothesis of Earthquakes," in *Rock Deformation*. New York, N. Y.: The Geological Society of America, Memoir 79.

GROBBELAAR, C., 1970, *A Theory of the Strength of Pillars*. Johannesburg: Voortrekkerpers.

KEHLE, R. O., 1964, "The Determination of Tectonic Stresses through Analysis of Hydraulic Well Fracturing," *Journal of Geophysical Research*, Voll. 69, pp. 259–273.

KOMARNITSKII, N. I., 1968, *Zones and Planes of Weakness in Rocks and Slope Stability*. New York, N. Y.: Consultants Bureau, p. 15.

KRSMANOVIĆ, D., 1967, "Initial and Residual Shear Strength of Hard Rocks," *Géotechnique* (London), Vol. 17, pp. 145–160.

LAMA, R. D., and V. S. VUTUKURI, 1978, *Handbook on Mechanical Properties of Rocks*, Vol. II, III and IV. Clausthal, Germany: Trans Tech Publications.

LOMNITZ, C., 1956, "Creep Measurement in Igneous Rocks," *Journal of Geology*, Vol. 64, No. 5, pp. 473–479.

MOHR, F., 1963, *Gebirgsmechanik*. Goslar: H. Hübner.

MÜLLER, L., 1965, "Safety in Rock Engineering" ("Die Sicherheit im Felsbau"), 15th Colloquium of the Austrian Regional Group of the International Society of Rock Mechanics, held September 24–25, 1964, at Salzburg. Wien: Springer.

OBERT, L., 1964, "In Situ Stresses in Rock, Rainier Mesa, Nevada Test Site, Operations Nougat and Storax,:" WT–1869, U.S. Bureau of Mines, College Park, Md.

PAVLOVIĆ, M., 1970, "Determining of Quasihomogeneous Zones of Elasticity and Deformability Characteristics of Rock Mass in Tunnel on the Basis of In Situ Investigations," *Proceedings Second Congress of the International Society for Rock Mechanics*, held in 1970 at Belgrade, Yugoslavia.

PERSEN, L. N., 1975, *Rock Dynamics and Geophysical Introduction to Stress Waves in Rocks*. Amsterdam–New York, N. Y.: Elsevier Scientific Publishing Company.

ROBERTS, A. F., 1977, *Geotechnology*. Oxford: Pergamon Press.

ROCHA, M., 1965, *Some Problems on Failure of Rock Mass*. Lisbon: Laborátorio Nacional de Engenharia Civil, Memória No. 258, p. 7.

RODRIGUES, F., P., 1970, "Anisotropy of Rocks: Most Probable Surfaces of the Ultimate Stresses and of the Moduli of Elasticity," *Proceedings Second Congress of the International Society for Rock Mechanics*, Belgrade, Yugoslavia.

ROLFE, S. T., and J. M. BARSON, 1977, *Fracture and Fatigue Control in Structures: Applications of Fracture Mechanics*. Englewood Cliffs, N. J.: Prentice-Hall, Inc.

SINGH, D. P., 1975, "A Study of Creep of Rock," *International Journal of Rock Mechanics and Mining Sciences*, Vol. 12, No. 9, pp. 271–275.

TER-STEPANIAN, G., 1966, "Types of Depth Creep of Slopes in Rock Masses," *Proceedings First Congress of the International Society for Rock Mechanics*, Lisbon Vol. 2, pp. 157–160.

PART 3

STRESSES IN ROCK ABOUT CIRCULAR UNDERGROUND OPENINGS

CHAPTER 6

STRESS FIELDS

6-1. Force Field

Any design and engineering analysis in rock is a two-phase process, namely:

1. Definition of the force field acting upon the construction material (such as soil or rock, for example), and
2. Determination of the reaction of the material to that force field.

Thus, in the design of an engineering structure, one deals with the analysis of stresses acting within the structural elements, as well as with the various properties of the structural materials.

A *force field* is a region of space or a medium where at every point at any time there acts a definite, coordinated force — the so-called field force. Thereby, the coordination of the field force to the points in space may be real or potential.

Examples of field force are:

1. Real field force: specific gravity; and
2. Potential field force: electrical field.

In statics, one limits oneself to stationary force fields:

$$X = x(x, y, z),$$
$$Y = y(x, y, z),$$
$$Z = z(x, y, z).$$

If the line of action of the field force for all points passes through a fixed center, one speaks, then, of a central force, viz., the central field.

A plane force field is expressed as

$$\left.\begin{aligned} X &= x(x, y) \\ Y &= y(x, y) \\ Z &= 0. \end{aligned}\right\} \quad \text{force plane}$$

Force (stress) fields may result from a variety of causes, one of which is the direct application of load.

6-2. Primary Stresses in Sound Rock

The cause of stresses in sound (virgin) undisturbed rock is the weight of the superimposed, or overburden rock, sometimes spoken of as the primary, or initial, or natural, or roof pressure of rock *en masse*. It varies in magnitude with depth from the ground surface. Near the surface, natural rock stresses are influenced by the weight of the rock, tectonic forces (caused by folding of the earth's crust), jointing, fractures, restraint against lateral expansion, and hard spots.

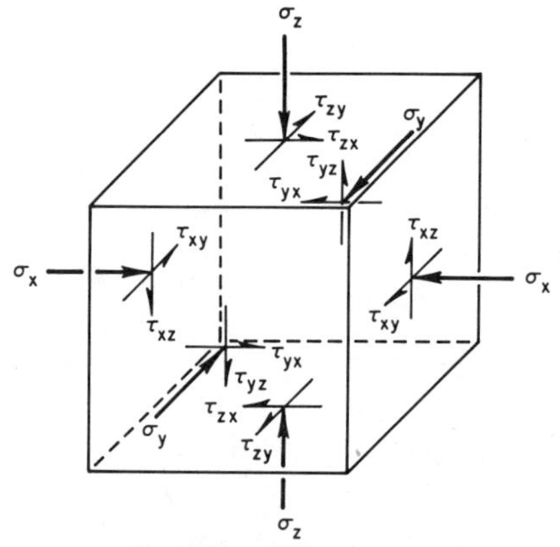

Fig. 6-1
Stress components in sound rock

Figure 6-1 illustrates general stress components in sound, homogeneous, isotropic rock. If the normal stresses are principal stresses, then the shear stresses vanish.

The magnitude and direction of the *initial* or primary stress in the rock — viz., the state of stress in the rock before excavation — depends upon the stress field. The state of stress in rock is disturbed or altered by excavations for underground openings and engineering structures. Excavations in rock bring about a new distribution of secondary or *induced* stresses within the rock around the excavations.

The amount and depth of this alteration are influenced by blasting and the local stress produced by the shape of the opening. According to test results as reported by TALOBRE (1957, Ref. 25 for Part 3), the depth over which the stresses are altered extends about one-half the width of the excavation. This assumes, of course, that the length of the excavation is several times greater than the width.

Assuming the rock is a homogeneous material, and knowing its exact elastic properties, the calculations of stresses in rock massifs, weakened by openings, can be made somewhat reasonably by means of the theory of elasticity.

It should be said, however, that in practice these assumptions are very seldom met.

Imagine an elementary cube. On this cube, there act the principal vertical stresses σ_z and the principal horizontal stresses. All these stresses act mutually perpendicularly.

Assuming that the rock mass is elastic and homogeneous (having in all directions the same physical-mechanical properties), both horizontal stresses σ_x and σ_y must be equal.

According to Professor HEIM's (5) hypothesis, rock stress at great depths is the same in all directions. Professor HEIM's theory, according to which the pressure at any point in the interior of a rock massif is the same in all directions (hydrostatic pressure distribution hypothesis), and the magnitude of the pressure is calculated as the pressure of the overburden rock ($\downarrow \sigma_v = \gamma \cdot z =$ geostatic pressure), may be regarded as a close approximation to real conditions in the rock massif at great depth z.

"Of all the assumptions as to the horizontal pressure, that of equal pressure in any direction is the most probable." (5)

It must be realized, though, that:

1. The usual depths in underground rock engineering are too small for HEIM's pressure distribution at "great depth".
2. Thermal and other factors may transform the rock into ta plastic state (gneiss, slate).
3. At usual depths, there is a difference between the magnitude of vertical and horizontal stresses, resulting in shear stresses.

4. Usually the largest stresses are not the vertical ones (by observation).
5. Stratification of rocks render inclined stresses.
6. Stresses in rock depend upon tectonic activity of the earth's crust (mountain-building forces).

Because the rock lateral confinement prevents the cube from expanding laterally under the action of the vertical stress σ_z, the sum of the strains in the elastic hemispace must be equal to zero:

$$\frac{1}{m} \cdot \frac{\sigma_z}{E} + \frac{1}{m} \cdot \frac{\sigma_y}{E} - \frac{\sigma_x}{E} = 0 \tag{6-1}$$

or

$$\sigma_z + \sigma_y - m \cdot \sigma_x = 0 \tag{6-2}$$

It follows that, if $\sigma_x = \sigma_y$, then

$$\sigma_x = \sigma_y = \frac{\sigma_z}{m - 1} = \frac{\mu}{1 - \mu} \cdot \sigma_z \tag{6-3}$$

where $m = 1/\mu$ = Poisson's number, an elastic property of a material. It is valid for the elastic interval of deformations. From seismic measurements the average value of m of the earth's crust is $m = 3.7$;

$\mu = 1/m$ = Poisson's ratio;

E = modulus of elasticity of rock as tested.

The modulus of elasticity E_c of the confined rock in situ may be ascertained as follows.

The vertical strain (compressive) is:

$$\varepsilon_v = \frac{p_v}{E_c} = \frac{p_v}{E} - \frac{2p_v}{Em(m - 1)} \tag{6-4}$$

Here E is the modulus of elasticity as obtained from an unconfined compression test, for example. Therefore:

$$E_c = E \frac{m(m - 1)}{(m + 1)(m - 2)} \tag{6-5}$$

For most rocks, Poisson's ratio μ is between 0.20 and $\mu = 0.33$, and $\mu/(1 - \mu) = \lambda_0$ should lie between 0.25 and 0.50.

For a compilation of Poisson's ratio for various kinds of soils and rocks, refer to Jumikis (13) and Jumikis and Jumikis (15).

TABLE 6-1

Lateral Pressure Coefficients λ_o for Various Types of Rocks

Rock	POISSON'S ratio μ	POISSON'S number m	$\lambda_o = \frac{\mu}{1-\mu} = \frac{1}{m-1}$
1	2	3	4
Diabase	0.33	3.00	0.500
Dolomite	0.083 – 0.20	12.00 – 5.00	0.091 – 0.250
Gabbro	0.125 – 0.20	8.00 – 5.00	0.143 – 0.250
Gneiss	0.11	3.30 – 6.60	0.303 – 0.152
Granite	0.15 – 0.24	0.66 – 4.16	0.177 – 0.312
Limestone	0.16 – 0.23	6.25 – 4.35	0.190 – 0.299
Limestone	0.14 – 0.17	7.00 – 6.00	0.170 – 0.200
Marble	0.25 – 0.38	4.00 – 2.63	0.333 – 0.613
Sandstone	0.17	5.88 – 2.80	0.205 – 0.556
Schist	0.08 – 0.20	12.50 – 5.00	0.087 – 0.250
Shale	0.11 – 0.54	9.10 – 1.85	0.123 – 1.176
Tuff	0.11	9.10	0.123

Table 6-1 shows computed λ_o-values of some common rock types. However, most of the in-situ measured values of $\sigma_h/\sigma_v = \lambda_o = \mu/(1-\mu)$ lie between 0.5 to 0.8 for hard rock, and between 0.8 and 1.0 for soft or inelastic rocks such as shale or salt (OBERT and DUVALL, 1967, Ref. 19).

If for limestone $m \approx 6\text{-}7$, then the horizontal stress σ_x is:

$$\sigma_x = (0.20)\,\sigma_z \text{ to } (0.17)\,\sigma_z$$

i.e., the horizontal stress σ is 17% to 20% of the vertical stress.

Assuming a gravitational stress field, the weight of the superimposed overburden in undisturbed condition exerts a vertical pressure σ_z at any point below the ground surface:

$$\sigma_z = \gamma \cdot h$$

where γ = unit weight of rock, and
h = thickness of rock layer.

From Eq. (6-3) one sees that the smaller m is, the larger the horizontal stress becomes.

It should be said here again that in order to specify completely a gravitational stress field, it must be assumed that:

1. The rock is linear-elastic, isotropic and homogeneous.
2. The lateral rock constraint or confinement is complete.
3. There are no stresses of tectonic origin such as those accompanying folding, shrinkage, or other distortions of the earth's crust.

In some instances, these tectonic stresses, together with the effects caused by inelasticity, inhomogeneity, and anisotropy, are known to affect the gravitational force field considerably. The rule that the largest principal stress in rock is generally acting vertically may be upset by tectonic stresses. For example, it has been reported that in the power plant cavern of the Snowy Mountain power scheme, Australia, the horizontal stresses were about 20% higher than the vertical stresses because of tectonic stresses.

6-3. In-Situ State of Stress

The in-situ state of vertical stress at a depth h, which depth is small as compared with the radius of the earth, is expressed as:

$$\sigma_z = \sigma_v = \gamma h \tag{6-6}$$

and the coordinated lateral stress is:

$$\sigma_x = \sigma_y = \sigma_h = [\mu/(1-\mu)] \cdot \sigma_v = \lambda_o \cdot \sigma_v \tag{6-7}$$

where $\sigma_v = \sigma_z =$ vertical, normal stress component;
$\sigma_h = \sigma_x = \sigma_y =$ horizontal, normal stress component;
$h =$ vertical depth below ground surface;

$$\lambda_o = \mu/(1-\mu) = \frac{1}{m-1} \tag{6-8}$$

$\sigma_h/\sigma_v =$ ratio of horizontal to vertical stress. It is a constant characterizing the stress field;
$\lambda_o \cdot \sigma_v =$ pressure at rest.

Here σ_v and σ_h are applied stresses.

In these discussions, compressive stresses are designated as positive and tensile stresses as negative.

Physico-geometrically, the stress components σ_x, σ_y and σ_z represent a stress ellipsoid.

When the lateral rock constraint is not completely rigid, λ_o will attain higher values than an ideally complete rigid constraint. However, the real value of λ_o for rock is difficult to determine (15).

6-4. Stress Fields

The stress field, sometimes referred to as the pre-existing stress, or initial or primary stress, or as the stress of the medium, is the state of stress that exists in the rock prior to mining or prior to any other kind of rock excavation or boring.

According to OBERT and DUVALL (19), three kinds of stress fields in rock may exist.

The three basic kinds of assumed stress fields are:

A *uniaxial* (unidirectional) *stress field,* a *biaxial* (two directional) *stress field,* and a *hydrostatic stress field.*

A uniaxial, unconstrained stress field is one where the medium (rock material) is subjected to compressive or tensile stresses in one direction only. Here $\lambda_0 = 0$. This state of stress would be encountered in rock at shallow depth below ground surface, and also near free vertical surfaces.

A biaxial stress field is one where the rock medium is subjected to compressive or tensile stresses in two mutually perpendicular directions. Here $\lambda_0 = 1/3$. This state of stress condition would be encountered at greater depth over a wide range of depth, depending upon type of rock, and at a depth of approximately 1000 m and more.

If in Eq. (6-8) one introduces $\mu = \dfrac{1}{m} = 1/4$, then $\lambda_0 = 1/3$. Such a state of stress pertains to the condition of a complete lateral constraint (no lateral deformation) in rock whose $\mu = 1/4$.

A hydrostatic stress field is one where the rock medium is subjected to equal stresses in three mutually perpendicular directions. Here $\lambda_0 = 1.0$. This means that σ_z, σ_x, and σ_y are all of equal magnitude of pressure. Such a condition when $\lambda_0 = 1.0$ may be encountered at great depth, and in semi-viscous or plastic rocks.

The stress fields are expressed analytically as:

$$\sigma_v = \gamma \cdot h$$
$$\sigma_h = \lambda_0 \cdot \sigma_v \qquad (6\text{-}9)$$

These three kinds of stress fields are sketched in Fig. 6-2.

Knowledge about the state of stress in rock is necessary in designing structures in rock and in evaluating the stability of underground structures.

HORVATH (1965, Ref. 6) put to the fore the idea that for a rock with certain characteristics of yield stress, unit weight, and POISSON's ratio, there

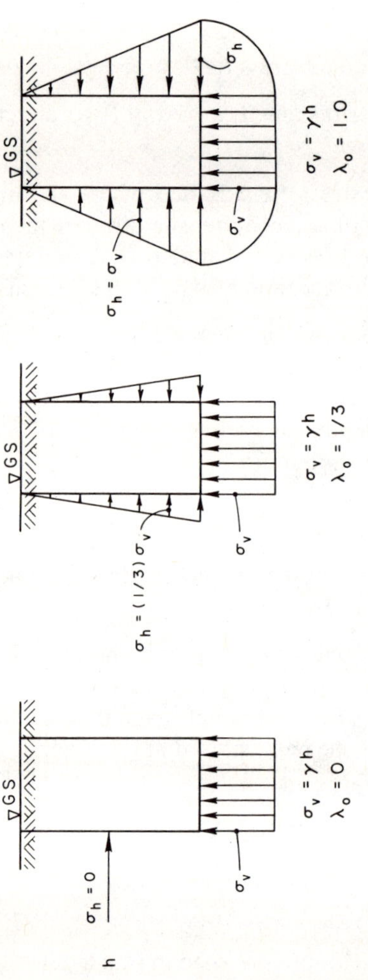

a) Uniaxial unconstrained stress field

b) Laterally constrained biaxial stress field

c) Hydrostatic stress field (in semiviscous or plastic rocks)

Fig. 6-2
Three assumed kinds of stress fields

is a limiting depth above which the lateral stresses may be computed on the basis of the theory of elasticity, but below which the horizontal principal stresses σ_2 and σ_3 may be derived from a plastic yield criterion, as:

$$\sigma_2 = \sigma_3 = \sigma_1 - \sigma_y = \gamma \cdot h - \sigma_y \quad (6\text{-}10)$$

where $\sigma_1 = \gamma h$, and
σ_y = yield stress.

Most underground structures are built at depths where the hydrostatic stress condition is not attained. Hence, for the purpose of stress analysis in rock around an opening (such as a tunnel, for example), it is necessary to know the lateral pressure coefficient at rest, λ_o. Otherwise, as is frequently done, λ_o is calculated by Eq. (6-8) by means of a known, or assumed, Poisson's number m for a particular rock.

Because of the existence of faults, joints, folds, stratification, anisotropy, and mechanical heterogeneity of rocks, the stresses in rocks can vary quite a lot from one point to another.

In rock mechanics and rock engineering, the engineer cannot choose a rock material with known mechanical properties and known initial state of stress because they are unknown.

Also, in rock engineering, usually the engineer has no wide latitude for selecting an ideal construction site.

There is no real justification for assuming that $\sigma_h = \lambda_o \cdot \sigma_v = [\mu/(1-\mu)](\gamma h)$. Any one geological event could have altered the horizontal stress considerably from the above value. During the geological past, the stress that exists today may have been acted upon by various geological processes. For example, in a tectonically disturbed area of active regional subsidence, the center of the area would undergo a compressive strain, while the periphery and stresses would be under tensile strain. Also, faulting, folding, and other geological processes would bring about certain states of stress, which stresses may differ considerably from those as evaluated by means of the theory of elasticity. Rheological processes in, and weathering of, rocks would alter the initial stress field also.

It must be realized that the state of stress in rock is the sum of forces and processes of various origins and natures.

Practically, a reliable stress field can be determined only by in-situ measurement. Hence, there is a need for determining quality indices of rocks by means of which the engineer could quickly and economically test, evaluate, and characterize the stress field in rock and the mechanical properties of a rock mass. These indices are needed for assessment of depths of foundations and underground openings, as well as for stability of rock slopes. In other words, they are important in every geotechnical undertaking.

CHAPTER 7

ELASTIC STRESS ANALYSIS IN ROCK AROUND UNDERGROUND OPENINGS

7-1. Underground Openings

One distinguishes between openings of horizontal underground structures with circular or elliptical cross-sections (tunnels, adits), vertical openings with circular cross-section (shafts, borings), and inclined tunnels (aqueous pressure penstocks in rock, for example).

Underground power plants and railroad yards (Oslo) are unaffected by topographic and climatic conditions such as freezing and avalanches. The surface landscape need not be destroyed. Such underground structures are also relatively more bombproof than surface generating stations.

The pressure tunnels may be unlined or lined with a shell. In unlined tunnels, the inherent strength of the rock is utilized.

Incompetent rock may be reinforced for the purpose of transmitting the lining pressure to the rock.

7-2. Secondary Stress Conditions in Rock

Generally, by an underground structure one refers to any excavated or natural subsurface opening, or system of openings, that is virtually self-supporting.

A self-supporting system is one in which the structural stresses are carried on the walls, pillars, and other unexcavated parts of the opening.

By the term "subsurface" or "underground" openings in rock is understood not merely adits, galleries and tunnels for vehicular traffic, aqueous pressure tunnels, and tunnels for river diversion, but also excavations in rock for shafts, subsurface factories, underground power plants, transformer chambers, garages, storage places, warehouses, shelters, arsenals, fortifications, and other possible uses. The purpose of tunnels is to provide for a) passageways for vehicular traffic, such as roads and railroads through mountains and under bodies of water; b) open and pressure conduits for water, and access to various underground spaces and structures.

7-3. Stresses in a Thick-Walled Cylinder

As an outset for stress analysis in rock, let us first avail ourselves to stresses in a thick-walled cylinder (16, 17, 19, 24, 27, 29).

Consider a circular, annular disk sliced out from a long, horizontal tunnel. The inside radius of the disk is r_i, its outside radius is r_o, and r is a variable radius (Fig. 7-1). The thick-walled cylinder is subjected to a uniform inside or internal pressure p_i, and to a uniform outside or external pressure p_o. These pressures induce in the wall, at any point, normal radial and normal tangential stresses, σ_r and σ_t, respectively.

The problem of determining the radial and tangential stresses at any point on a thick-walled cylinder, as a function of the applied pressures and radii involved, was solved by the French elastician GABRIEL LAMÉ in 1833. The elements of a thick-walled cylinder are shown in Fig. 7-1. The thick wall may be thought of as composed of thin, annular shells.

The radial stresses in the shell are designed as σ_r. The tangential stresses in the shell are designated as σ_t. Compressive stresses are indicated as positive. Tensile stresses are designated as negative.

A relationship between the radial and tangential stresses can be obtained from the equilibrium condition by summation of projections of all forces along a certain axis. The summation of forces must be equal to zero.

Referring to Fig. 7-1b, set $\Sigma F_r = 0$ along radius r to the volume element dV:

$$\Sigma F_r = \sigma_r \cdot r dw - \left(\sigma_r + \frac{d\sigma_r}{dr} \cdot dr\right)\left(r + dr\right) dw + 2 \cdot \sigma_t \cdot dr \cdot \frac{dw}{2} \qquad (7\text{-}1)$$

$$-\sigma_r \cdot dr - d\sigma_r \cdot r - \sigma_t \cdot dr = 0$$

$$\sigma_r + r \cdot \frac{d\sigma_r}{dr} = \sigma_t$$

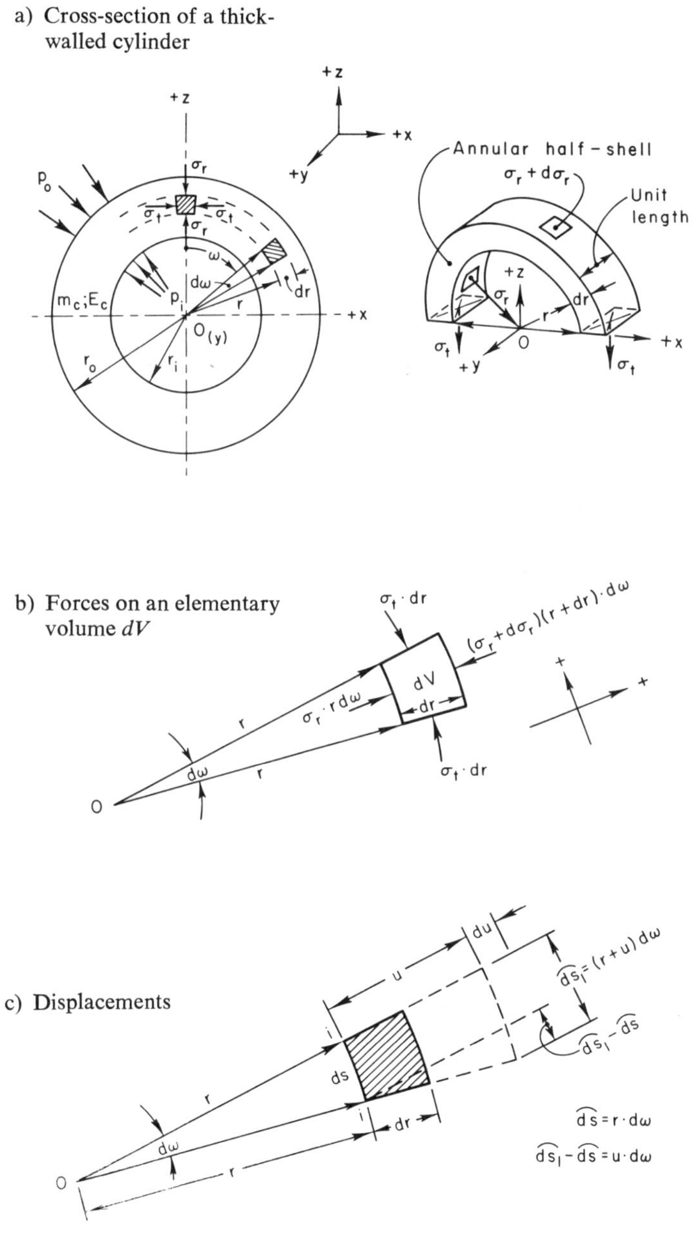

Fig. 7-1
Elements of a thick-walled cylinder

or

$$\sigma_t - \sigma_r - r \cdot \frac{d\sigma_r}{dr} = 0. \tag{7-2}$$

This is the equilibrium equation for the volume element dV.

Equation (7-2) can also be written in a differential form:

$$\sigma_t = \frac{d}{dr}(r \cdot \sigma_r) \tag{7-3}$$

Another relationship between radial and tangential stresses is obtained from the assumption that plane cross-sections remain plane, and hence the longitudinal strain ε_y is constant for all fibers.

Applying the extended HOOKE's law for triaxial stress (11), based on the plane stress condition, the longitudinal strain ε_y is written as:

$$\varepsilon_y = \frac{1}{E}\left[\sigma_y - \frac{1}{m}\left(\sigma_r + \sigma_t\right)\right] \tag{7-4}$$

where $m = 1/\mu$ = POISSON's number and $\mu = 1/m$ = POISSON's ratio. Because m, E, ε_y, and σ_y are all constant quantities:

$$\sigma_r + \sigma_t = \text{const} \tag{7-5}$$

throughout the cross-section of the thick-walled cylinder.

If $\varepsilon_y = 0$, then:

$$\sigma_y = \frac{1}{m}\left(\sigma_r + \sigma_t\right) \tag{7-6}$$

The elasticity equations for the plane stress condition (remember the long cylinder) expressing the other strains are:

$$\varepsilon_r = \frac{1}{E}\left(\sigma_r - \frac{1}{m} \cdot \sigma_t\right) \tag{7-7}$$

$$\varepsilon_t = \frac{1}{E}\left(\sigma_t - \frac{1}{m} \cdot \sigma_r\right) \tag{7-8}$$

From the geometry (Fig. 7-1c), the interior side i-i of the elementary volume is displaced by an amount u, and the outer one by $u + du$. Thus, the original thickness dr of the annular, elementary shell changes by an amount du. The elastic, radial strain component is:

$$\varepsilon_r = \frac{du}{dr} \tag{7-9}$$

The elastic tangential strain component ε_t in a symmetrical system in the tangential direction is computed as follows:

$$\varepsilon_t = \frac{ds_1 - ds}{ds} = \frac{(r + u)\,dw - rdw}{rdw} = \frac{u}{r} \tag{7-10}$$

$$\therefore \quad u = r \cdot \varepsilon_t \tag{7-11}$$

$$\therefore \quad \varepsilon_r = \frac{du}{dr} = \frac{\partial\,(r \cdot \varepsilon_t)}{\partial r} = \varepsilon_t + r \cdot \frac{\partial \varepsilon_t}{\partial r} \tag{7-12}$$

$$\therefore \quad \varepsilon_r - \varepsilon_t - r\frac{\partial \varepsilon_t}{\partial r} = 0. \tag{7-13}$$

From Equations (7-7), (7-8), (7-9) and (7-10), obtain the radial and tangential stress components as

$$\sigma_r = E \cdot \frac{m}{m^2 - 1}\left(m \cdot \frac{du}{dr} + \frac{u}{r}\right) \tag{7-14}$$

$$\sigma_t = E \cdot \frac{m}{m^2 - 1}\left(\frac{du}{dr} + m \cdot \frac{u}{r}\right) \tag{7-15}$$

Substitution of Eqs. (7-14) and (15) into Eq. (7-2) renders the Euler's differential equation for the radial displacement u:

$$\frac{du}{dr} - \frac{u}{r} - r \cdot \frac{d\left(\frac{u}{r}\right)}{dr} = 0$$

or

$$r^2 \cdot \frac{d^2u}{dr^2} + r \cdot \frac{du}{dr} - u = 0 \tag{7-16}$$

With $u = f\,(r)$, the elastic strain components ε_r and ε_t, as well as the corresponding elastic normal stress components σ_r and σ_t, can be calculated. Therefore, our immediate task is to determine the expression of $u = f\,(r)$.

Let the outset for the solution of Euler's homogeneous, linear differential equation of the second order be:

$$u = A \cdot r + \frac{B}{r} \tag{7-17}$$

where A and B are constants to be determined.

Divide by r:

$$\frac{u}{r} = A + \frac{B}{r^2} \tag{7-18}$$

Differentiate u (Eq. 7-17) with respect to r:

$$\frac{du}{dr} = A - \frac{B}{r^2} \tag{7-19}$$

The boundary conditions are:

1) when $r = r_i$, $\sigma_r = p_i$
2) when $r = r_o$, $\sigma_r = p_o$

To determine the stress components as a function of r, substitute u- and (u/r)-values from Eqs. (7-17) and (7-18) into Eqs. (7-14) and (7-15) to obtain

$$\sigma_r = E \cdot \frac{m}{m^2 - 1}\left[A\,(m+1) - \frac{B}{r^2}(m-1)\right] \tag{7-20}$$

$$\sigma_t = E \cdot \frac{m}{m^2 - 1}\left[A\,(m+1) + \frac{B}{r^2}(m-1)\right] \tag{7-21}$$

The constants A and B will now be determined from Eqs. (7-20) and (7-21) by means of the known boundary conditions:

$$A = \frac{m-1}{m \cdot E} \cdot \frac{1}{a^2 - 1}(p_o\, a^2 - p_i) \tag{7-22}$$

$$B = \frac{m+1}{m \cdot E} \cdot \frac{r_o^2}{a^2 - 1}(p_o - p_i) \tag{7-23}$$

where

$$a = \frac{r_o}{r_i} \tag{7-24}$$

With the known constants A and B, the radial displacement u of any point on the thick-walled cylinder, by Eq. (7-17), is:

$$u = \frac{1}{E} \cdot \frac{1}{a^2 - 1} \cdot \left[\frac{m-1}{m}(p_o \cdot a^2 - p_i)\, r + \frac{1}{r} \cdot \frac{m+1}{m} \cdot \frac{r_o^2}{a^2 - 1}(p_o - p_i)\right] \tag{7-25}$$

By setting in Eq . (7-25), $r = r_i$ and $r = r_o$, the displacements of the inner and outer contour of the disk, u_i and u_o, respectively, can be calculated.

With the known values of A and B, LAMÉ elastic stresses are:

$$\sigma_r = p_o \frac{a^2 - \alpha^2}{a^2 - 1} + p_i \frac{\alpha^2 - 1}{a^2 - 1} \text{ (compression)} \tag{7-26}$$

$$\sigma_t = p_o \frac{a^2 + \alpha^2}{a^2 - 1} - p_i \frac{\alpha^2 + 1}{a^2 - 1} \tag{7-27}$$

Here $a = r_o/r_i$, and

$$\alpha = r_o/r \tag{7-28}$$

Notice, that for the inner contour, where $r = r_i$:

$$\alpha = r_o/r_i = a \tag{7-29}$$

and

$$\sigma_{ri} = p_i \tag{7-30}$$

$$\sigma_{ti} = p_o \frac{2a^2}{a^2 - 1} - p_i \frac{a^2 + 1}{a^2 - 1} \tag{7-31}$$

For the outer contour of the thick-walled cylinder:

$$\alpha = r_o/r_o = 1 \tag{7-32}$$

$$\sigma_{ro} = p_o \tag{7-33}$$

$$\sigma_{to} = p_o \frac{a^2 + 1}{a^2 - 1} - p_i \frac{2}{a^2 - 1} \tag{7-34}$$

With $r_o \to \infty$, and for inside pressure p_i only (i. e., when $p_o = 0$), by Eqs. (7-26) and (7-27):

$$\sigma_r = p_i \frac{r_i^2}{r^2} \text{ (C)} \tag{7-35}$$

$$\sigma_t = - p_i \frac{r_i^2}{r^2} \text{ (T)} \tag{7-36}$$

On the tunnel contour, where $r = r_i$:

$$\sigma_{r_{max}} = p_i \tag{7-37}$$

$$\sigma_{t_{min}} = -p_i \tag{7-38}$$

When $r = r_i$, the strain ε_t in tangential direction is:

$$\varepsilon_t = \frac{2\pi(r + \Delta r) - 2\pi r}{2\pi r} = \frac{\Delta r}{r} \tag{7-39}$$

$$\therefore \quad \Delta r = \varepsilon_t \cdot r_i = -\frac{1}{E_{rock}} \cdot \left(\frac{m+1}{m}\right) \cdot p_i \cdot r_i \tag{7-40}$$

where m is the POISSON's number for rock.

7-4. Theoretical Basis for Stress Analysis in Rock

In a plane stress condition, all stresses induced in rock around a circular opening (such as a tunnel) act in the cross-sectional planes of the long structure. Perpendicular to the cross-sectional slice areas, no stresses are transferred.

At this point, it should be remembered from Section 6-3, that the primary stresses, such as $\sigma_z = \gamma h$ and $\sigma_h = \sigma_x = \sigma_y = [1/(m-1)]\ \sigma_r = \lambda_o \cdot \sigma_r$, are the primary stresses in an ideal, undisturbed medium of rock.

However, the stress field in a rock changes by the presence of an underground opening within the rock. Also, a stress concentration occurs in the rock around the opening.

After excavating a tunnel or making any other cavity in the rock, vertical and horizontal forces around the opening are set free.

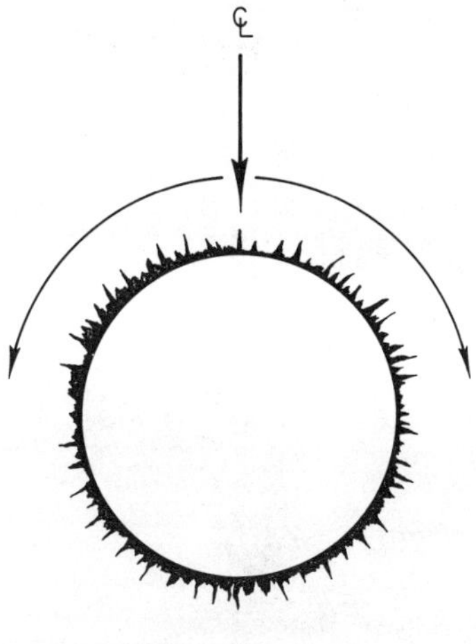

Fig. 7-2
By-passing pressure

These forces, prior to excavation, were carried by the rock. The pressure of the overburden rock must now, after excavation, bypass the opening in the rock, and must be transferred through the adjacent rock (Fig. 7-2). The rock around the opening (say tunnel) loses its support, and can now imperceptibly begin to deform.

An underground opening brings about a lateral squeeze of vertical force lines (Fig. 7-3). This results in a large stress concentration at the abutments *a* and *b* (springings of a vault). Also, one should be aware that between the tunnel vault and the natural arching of rock tension zones form at the crown and invert (at the top and bottom of the tunnel) of the vault.

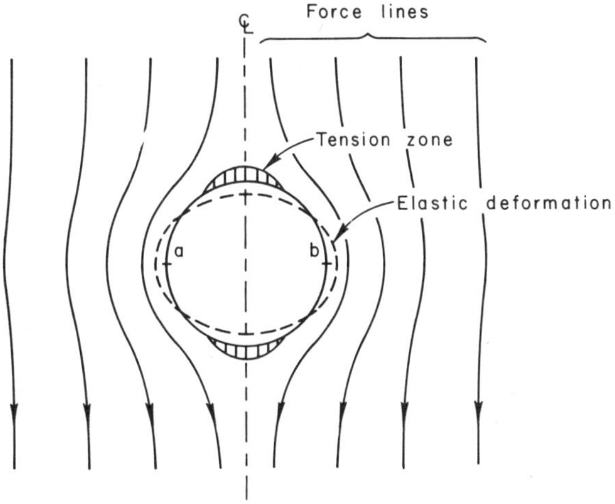

Fig. 7-3
Concentration of force lines, elastic deformation, and tension zones around a circular underground opening

Thus, the stress condition around the opening changes considerably, resulting here in the so-called induced, secondary stress condition.

Upon transition from the primary to the secondary stress condition, elastic and plastic deformations of rock occur.

The reasons for inducing secondary stress conditions in rock are:

1. Loosening of the rock mass around cavities;
2. Weight of the overburden rock mass;
3. Tectonic forces, and

4. Volumetric expansion of the rock mass by thermal effects, or by swelling brought about by the action of physical or physico-chemical processes.

Depending upon the kind and properties of the rock, secondary stress conditions require the use of different construction methods. All of the above-mentioned four secondary stress (pressure) conditions may also occur simultaneously.

Relative to probable secondary stress conditions in rock, they may set in:

1. In sound rock;
2. In pseudo-solid, soft, and weathered rock; and
3. In loose rock, viz., soil.

Sound rock transfers externally applied load by beam action limited by beam deflection. In loose rocks and in soil, load transfer takes place, most likely, by friction that develops during mass displacement. In order to maintain the underground opening, the intrusion of the rock into the opening must be prevented by rock-supporting structures. The load acting upon the supports is referred to as rock pressure. Remember: *"Ut tensio, sic vis."* This old principle has also been recognized at the First International Congress on Rock Mechanics in 1966 held in Lisbon, Portugal.

Stresses in rocks that are weakened by subsurface openings, which stresses remain below the elastic limit, in special cases may be calculated theoretically by means of the theory of elasticity.

The theoretical analysis of elastic, plane stress conditions in a circular plate of infinite extent, in which plate there prevailed an assumed homogeneous stress condition prior to introducing a circular opening, is a relatively simple one (7, 29). This assumption, in reality, applies to the rock, the better, the deeper is the opening below the ground surface. This may be so because at great depth, the consideration of stress increase vertically by the force of the rock mass may be disregarded in the region under consideration.

The opinion that inhomogeneity and anisotropy are insignificant at large pressures of the rock mass does not apply, because upon excavating an underground opening there sets in reduction or *relaxation* of stresses in the rock mass.

Assuming that:

1. The rock mass is an ideally elastic, homogeneous, and isotropic medium;
2. Hooke's law between stress and strain is valid;

3. The thickness of rock overburden is large as compared with the diameter of the tunnel, so that σ_v and σ_h can be considered to be constant within the influence zone of the tunnel, i. e., $\sigma_v = \gamma_h$, and $\sigma_h = \lambda_o \cdot \gamma \cdot h$;

the secondary stress conditions in rock can be computed by means of theory of elasticity.

In a very hard rock, with no planes of separation, the stress condition, as calculated, is approximately true, and the stress distribution equals the theoretical stress distribution within the elastic, homogeneous, and isotropic medium. These assumptions, however, do not apply in layered and schistose rocks.

In studying stresses in polar systems such as horizontal tunnels with circular cross-section, it is convenient to perform calculations in the polar, or cylindrical, coordinate system.

As a start of analysis, assume now a thick-walled, horizontal tunnel from which a vertical slice perpendicular to the tunnel axis is sliced off at a point that is far from both ends of the tunnel.

Now let the outer radius r_o of the cylinder approach infinity. Thus, the rock boundaries around the opening are infinite, representing the case of stress condition in rock about a freshly excavated, unlined tunnel or underground opening (Fig. 7-4).

The stresses acting perpendicularly to the cross-sectional plane are usually not of much interest in the secondary stress analysis of circular tunnels in rock. Therefore, for purpose of simplicity, tunnel problems in rock are analyzed under the assumption of the state of plane stress.

Obviously, after the fresh excavation of the opening, the primary stresses undergo alteration in the vicinity of the tunnel circumference, because the rock tends to fill the opening as soon as the driving and excavation of the tunnel commence.

Because the deformations of the slice in the direction of the longitudinal axis of the horizontal tunnel are counteracted in the same measure presumably by the deformations of the adjacent rock slices in contact with the slice under consideration, the stress analysis can be based on the mechanics of a circular disk by means of elasticity equations (LAMÉ equations for a thick-walled cylinder).

Calculations of stresses in a rock massif weakened by openings and cavities require the use of the theory of elasticity and the theory of plasticity.

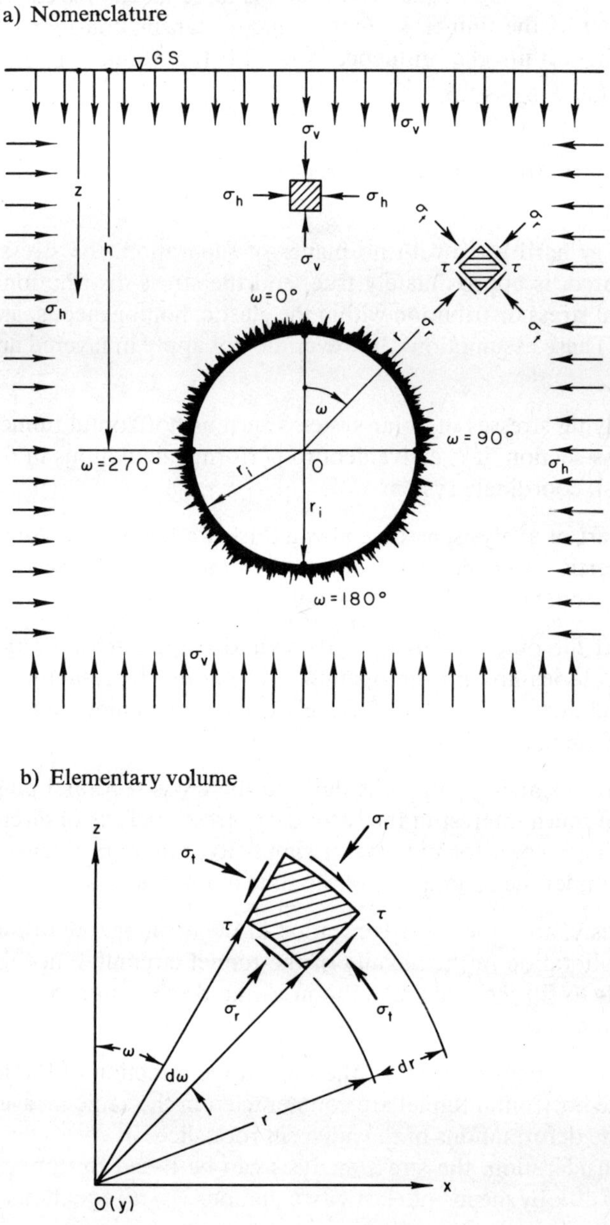

Fig. 7-4
Tunnel-rock system in a biaxial stress field;
σ_v and σ_h are primary stresses

The radial, tangential, and shear stresses are given by the theory of elasticity (19, 29) as:

$$\sigma_r = \frac{\sigma_v}{2}\left[(1+\lambda_o)(1-a^2)+(1-\lambda_o)(1+3a^4-4a^2)\cdot\cos 2\omega\right] \tag{7-41}$$

$$\sigma_t = \frac{\sigma_v}{2}\left[(1+\lambda_o)(1+a^2)-(1-\lambda_o)(1+3a^4)\cdot\cos 2\omega\right] \tag{7-42}$$

$$\tau_{rt} = -\frac{\sigma_v}{2}(1-\lambda_o)(1-3a^4+2a^2)\cdot\sin 2\omega \tag{7-43}$$

where $\sigma_v = \gamma \cdot h$ = vertical overburden stress,
γ = unit weight of rock,
h = thickness of overburden,
r_i = inside radius of opening,
r = variable radius,
ω = amplitude = angle between vertical (z) axis and radius r,
$\lambda_o = \dfrac{1}{m-1} = \dfrac{\mu}{1-\mu} = \sigma_h/\sigma_v$ = rock lateral pressure coefficient at rest,
$a = r_i/r$ for brevity. (7-44)

Equations (7-41) to (7-43) show that the extreme values of these stresses around the contour of the tunnel opening are determined by the cos 2ω and sin 2ω functions. Hence, the stress diagrams about the horizontal and vertical axes must be symmetrical. Especially informative for the stability of the tunnel perimeter are the extreme tangential stresses.

The extreme values of the tangential stress along the perimeter ($r = r_i$) of the circular tunnel opening can be determined by setting $d\sigma_t/d\omega = 0$

1. When $\omega = 0$ and $\omega = \pi$ (crest and invert):

$$\sigma_{t\,min} = -\sigma_v(1-3\lambda_o) \tag{7-45}$$

i. e., tension when $\lambda_o < \dfrac{1}{3}$

2. When $\omega = \pm\,\pi/2$:

$$\sigma_{t\,max} = \sigma_v(3-\lambda_o) \tag{7-46}$$

Depending upon the magnitude of λ_o, these calculations render along the vertical and horizontal direction (along the x- and z-axes) the course of stresses as shown in Fig. 7-5.

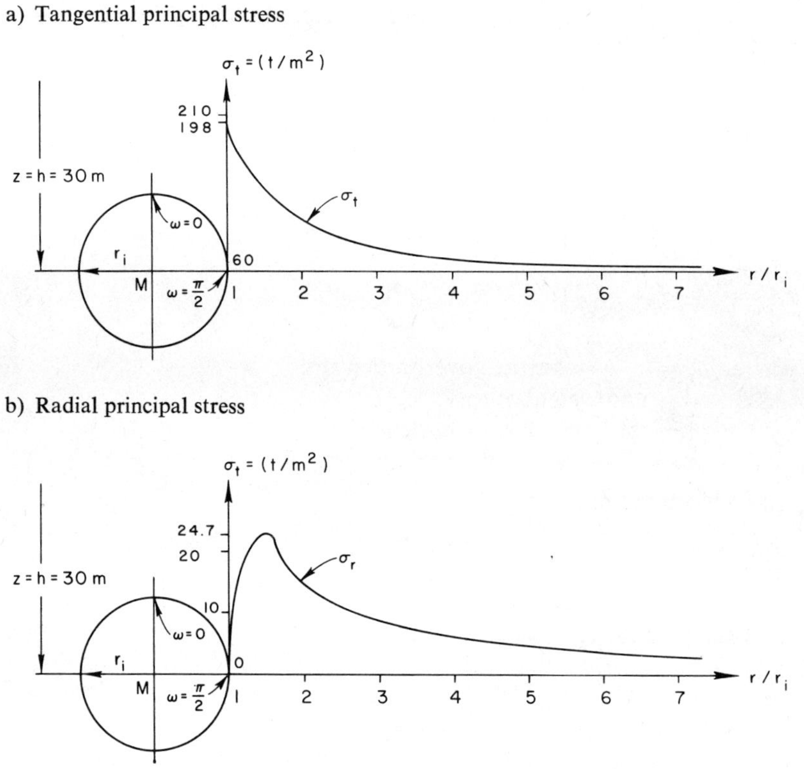

Fig. 7-5
Stress distribution diagrams around a circular opening in rock for $\gamma = 2.2$ t_f/m^3, $h = 30$ m, and $\omega = 90°$

Theoretically, the factor λ_o varies from $\lambda_o = 0$ to $\lambda_o = 1.0$ (see Chapter 6 on stress fields).

Typical stress distribution around circular openings in rock are shown in Figs. 7-5a and 7-5b. In each case, especially high stresses prevail along the periphery or contour of the tunnel. σ_t stresses at the springings ($\omega = 90$ °) attain two to three times the amount of the average vertical stresses σ_v. For $\omega = 90°$, $\tau_{rt} = 0$ (by Eq. 7-43).

If the horizontal stress is of the same magnitude as the vertical stress, then all stresses at any point around the opening are symmetrical, i. e., a compressive stress of magnitude of $2\sigma_v$. If there is no horizontal stress, then $\sigma_h = 0$ and $\lambda_o = 0$. In such a case, there may occur considerable tensile stresses in a zone between the tunnel vault and the natural arching of rock;

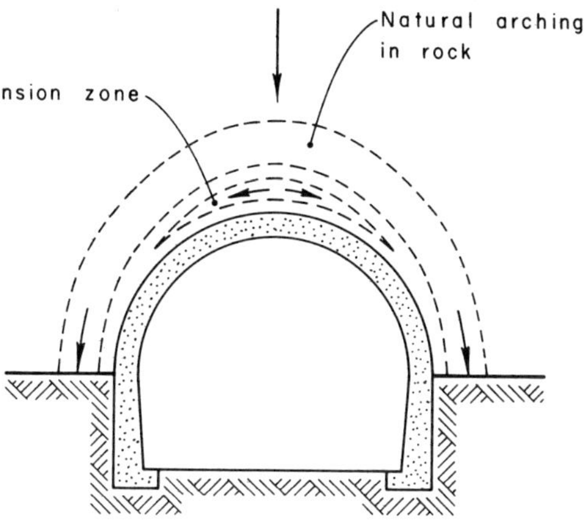

Fig. 7-6
Tension zone above an excavated underground opening below the natural arching

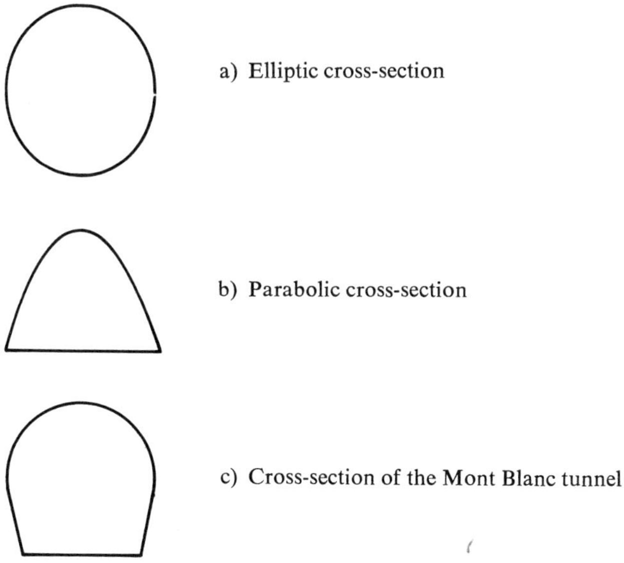

Fig. 7-7
Some cross-sections of underground openings

at the roof point the magnitude of these tensile stresses may attain the value of $-\sigma_v$.

Between these two extreme values, there is an intermediate case where the magnitude of the stresses at the roof point ($\omega = 0$) has a value of zero. This occurs when $\lambda_o = 1/3$, i. e., when the horizontal stress has a value of one-third of that of the vertical stress, corresponding to a POISSON's ratio of $\mu = 0.25$, or $m = 4.0$. A lateral deformation which produces such a high stress may be brought about by a massive granite, basalt, diorite, augite, or limestone. For all other rocks, m is less than 4.0; hence, there exists frequently above the vertix (crown) a zone of tension. This tension zone above a vault of the circularly arched tunnel frequently brings about cracks in the rock, so that the blocks separated from one another above the crown lose their lateral support and are free-hanging (Fig. 7-6). Such free-hanging rock fragments are very dangerous, and should be avoided by choosing a cross-section other than circular (Fig. 7-7).

7-5. Specialization of Stress Equations

Of particular interest in rock mechanics is the dependence of the secondary stress condition as a function of λ_o.

a) $\lambda_o = 0$. This case represents the case of a uniaxial, primary stress condition. When $\lambda_o = 0$, Eqs. (7-41), (7-42), and (7-43) take the form:

$$\sigma_r = \frac{\sigma_v}{2}\left[(1 - a^2) + (1 - 4a^2 + 3a^4)\cos 2\omega\right] \tag{7-47}$$

$$\sigma_t = \frac{\sigma_v}{2}\left[(1 + a^2) - (1 + 3a^4)\cos 2\omega\right] \tag{7-48}$$

$$\tau = \tau_{rt} = +\frac{\sigma_v}{2}(3a^4 - 2a^2 - 1)\sin 2\omega \tag{7-49}$$

When $\omega = 90°$ and $\omega = 270°$, then at the rim of the opening where $a = r_i/r = 1.0$, there is an increase in the tangential stress up to the value of $\sigma_t = 3\sigma_v$. Thus, for $a = 1.0$:

$$\left.\begin{aligned} \sigma_r &= 0 \\ \sigma_t &= 3\sigma_v \\ \tau &= 0 \end{aligned}\right\} \tag{7-50}$$

When $\omega = 0$ (at the crown), the prevailing stresses in rock are:

$$\left.\begin{aligned} \sigma_r &= 0 \\ \sigma_t &= -\sigma_v \\ \tau &= 0 \end{aligned}\right\} \qquad (7\text{-}51)$$

b) $\lambda_o = 1/3$. If there should be no tensile stresses at the crown, i. e., $\sigma_t = 0$, then for $\omega = 0$, $r = r_i$, and $a = 1$, the following stress equation results:

$$\sigma_t = \frac{\sigma_v}{2}\left[2\,(1 + \lambda_o) - 4\,(1 - \lambda_o)\right] = 0 \qquad (7\text{-}52)$$

From this equation for the assumption as before, $\lambda_o = 1/3$.

When $\lambda_o = 1/3$, then $m = 4$; i. e., when POISSON's number $m \leqq 4$, there are no tensile stresses at the crown. For these conditions, when $\lambda_o = 1/3$ and $\sigma_t = 0$, the set of stress equations is:

$$\sigma_r = \frac{\sigma_v}{3}\left[2\,(1 - a^2) + (1 - 4a^2 + 3a^4)\cos 2\omega\right] \qquad (7\text{-}53)$$

$$\sigma_t = \frac{\sigma_v}{3}\left[2\,(1 + a^2) - (1 + 3a^4)\cos 2\omega\right] \qquad (7\text{-}54)$$

$$\tau = \frac{\sigma_v}{3}\,[3a^4 - 2a^2 - 1]\sin 2\omega \qquad (7\text{-}55)$$

c) $\lambda_o = 1.0$. In this case, $\sigma = \sigma_v = \sigma_h$, and the stresses are:

$$\sigma_r = \sigma\,(1 - a^2) \qquad (7\text{-}56)$$

$$\sigma_t = \sigma\,(1 + a^2) \qquad (7\text{-}57)$$

$$\tau = 0 \qquad (7\text{-}58)$$

At the rim, or contour of the opening, $a = 1$, and:

$$\sigma_r = 0 \qquad \sigma_t = 2\sigma_v \qquad \tau = 0 \qquad (7\text{-}59)$$

The largest and least values of the σ_r stresses are at the position where $\omega = 0°$ and $\omega = 90°$.

Hence, this specialization (c) of Eqs. (7-41) and (7-42) for $\lambda_o = 1.0$ and $a = r_i/r = 1.0$ at the contour shows that the value of the radial stress σ_r is equal to zero, and that the value of the tangential stress σ_t attains a value twice the original value of the primary stress, i. e.:

$$\sigma_t = \sigma_v\left[1 + \left(\frac{r_i}{r}\right)^2\right] = 2\cdot\sigma_v = 2\cdot(\gamma h) \qquad (7\text{-}60)$$

Figure 7-8 shows the theoretical distribution of the σ_r- and σ_t-stresses after the excavation of a circular opening in elastic rock. The asymptote for both stresses (σ_r and σ_t) is the same—namely, γh.

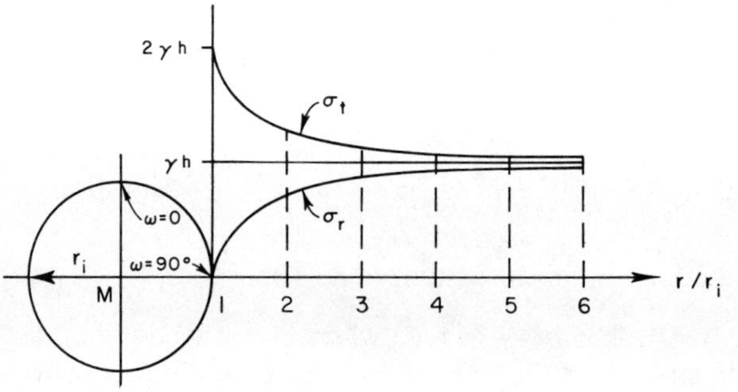

Fig. 7-8
Stress diagrams around an unlined tunnel for $\lambda_0 = 1.0$ at $\omega = 90°$

From this figure, it can be observed that beyond the abscissa whose value is $r/r_i > 4$, the stress increment induced is less than approximately 6 %.

Hence, practically, the extent of the disturbed zone may be taken as proximately twice the width of the opening.

The softer and the more fissured or plastically yielding the rock material is, the farther the maximum stress moves from the contour of the opening into the rock.

The biaxial stress component equations show that the stresses around the opening are independent of the elastic constants of the rock material, and of the radius of the opening. The radius enters into these equations only as a dimensionless ratio r_i/r, which specifies the distance from the boundary or contour of the opening.

7-6. Discussion of Equations

1. The magnitude of the stresses depends upon the depth of the opening, viz., upon σ_v, as well as upon Poisson's number m

$$[\lambda_o = 1/(m - 1)] \tag{7-61}$$

2. By equations (7-41) and (7-42), and Fig. 7-4, the largest and the smallest values of σ_r and σ_t occur in the vertical and horizontal sections,

respectively, where $\omega = 0°$ and $\omega = 90°$:

$$\sigma_{r_{\omega=0°}} = \frac{\sigma_V}{2}\left[(1+\lambda_o)(1-a^2)+(1-\lambda_o)(1+3a^4-4a^2)\right] \quad (7\text{-}41\,a)$$

$$\sigma_{t_{\omega=0°}} = \frac{\sigma_V}{2}\left[(1+\lambda_o)(1+a^2)-(1-\lambda_o)(1+3a^4)\right] \quad (7\text{-}42\,a)$$

and

$$\sigma_{r_{\omega=90°}} = \frac{\sigma_V}{2}\left[(1+\lambda_o)(1-a^2)-(1-\lambda_o)(1+3a^4-4a^2)\right] \quad (7\text{-}41\,b)$$

$$\sigma_{t_{\omega=90°}} = \frac{\sigma_V}{2}\left[(1+\lambda_o)(1+a^2)+(1-\lambda_o)(1+3a^4)\right] \quad (7\text{-}42\,b)$$

3. The shear stresses are at a maximum when $\omega = 45°$; by Eq. (7-43):

$$\tau_{rt_{max}} = -\frac{\sigma_V}{2}(1-\lambda_o)(1-3a^4+2a^2) \quad (7\text{-}43\,a)$$

4. At the contour of the circular opening where $r = r_i$, the radial stresses σ_r have a value equal to zero.

5. At the contour of the circular opening, where $r = r_i$, the tangential normal stresses σ_t assume the following values:

a) at the crown ($\omega = 0°$):

$$\sigma_{t_{\omega=0°}} = -\sigma_V(1-3\lambda_o) \qquad [T] \quad (7\text{-}42\,c)$$

b) at the abutment ($\omega = 90°$):

$$\sigma_{t_{\omega=90°}} = \sigma_V(3-\lambda_o) \qquad [C] \quad (7\text{-}42\,d)$$

i. e., the tangential stress at the abutment is always a compressive stress.

6. At the contour of the circular opening, the tangential stress (σ_t) is independent of the diameter of the opening. This means that by decreasing the diameter of the opening, the stresses at the abutment cannot be reduced.

7. The tangential stress σ_t along the contour of the circular opening is not constant. When POISSON's number $m = 4$, the stress at the crown is equal to σ_t = zero, and at the abutment it is

$$\sigma_t = (2.66)\,\sigma_V. \quad (7\text{-}62)$$

8. If $m < 4$, then at the crown there prevails a compressive tangential stress. If $m > 4$, then there prevails a tensile stress.

9. The tangential stress σ_t at the abutment is always a compressive stress.

10. When $m = 3$, $m = 4$, $m = 5$, and $m = 7$, the tangential stresses σ_t at the circular contour as shown in Table 7-1 result.

11. Between the crown and the abutment there is a place where the tangential stress $\sigma_t = 0$. Its position depends upon POISSON's number m. The angle $_{=0}$ is calculated by Eq. (7-42), introducing $r = r_i$ and setting the equation equal to zero to obtain:

$$\cos 2\omega = \frac{1 + \lambda_o}{2\,(1 - \lambda_o)} = \frac{m}{2\,(m - 2)} \tag{7-63}$$

12. Commonly, the tensile strength of a rock is considerably less than its compressive strength. Frequently it is:

$$\sigma_{\text{tensile}} \approx (0.1)\,\sigma_{\text{compressive}} \tag{7-64}$$

13. Also, it has been generally observed in practice that the larger the POISSON's number, the less is the tensile strength. This observation may give one an idea for choosing the necessary method of construction in rock relative to the kind of support (propping, anchor bolting, and the like).

14. The ratio of any one of the stress components at a point to one of the applied stresses is termed the stress concentration. The maximum positive stress concentration and the minimum negative stress concentration are called *critical stress concentrations* as defined by OBERT and DUVALL (19).

TABLE 7-1

Tangential Stresses σ_t at Contour of a Circular, Horizontal Tunnel in an Elastic Rock Mass

POISSON's Number m	Tangential stresses, σ_t: At the crown, $\omega = 0°$	Tangential stresses, σ_t: At the abutment, $\omega = 90°$
1	2	3
3	$+\,0.50 \cdot \sigma_V$ [C]	$2.50 \cdot \sigma_V$
4	0	$2.66 \cdot \sigma_V$
5	$-\,0.25 \cdot \sigma_V$ [T]	$2.75 \cdot \sigma_V$
7	$-\,0.50 \cdot \sigma_V$ [T]	$2.83 \cdot \sigma_V$

7-7. Elastic Deformations in Sound Rock Upon Excavation of a Cavity

Stress is always associated with an accompanying relative deformation of the material called the strain.

Upon excavation of a large cavity in rock, there takes place a transition from a primary stress condition in rock to a secondary stress condition, accompanied by relatively large elastic deformations of the rock material. This is so because upon making large openings in the rock, the process of a complete stress relaxation requires a long time.

In the deformation analysis that follows, assume again:

$\sigma_v = $ primary stress
$\sigma_h = \lambda_o \cdot \sigma_v = $ horizontal stress

Further, assume that the thickness h of the overburden is large as compared with the radius of the cavity $(h \gg r)$.

In the plane stress problem when $\omega = 0$, the secondary stresses by Eqs. (7-41), (7-42), and (7-43) are:

$$\sigma_r = \frac{\sigma_v}{2}\left[(1 + \lambda_o)(1 - a^2) + (1 - \lambda_o)(1 + 3a^4 - 4a^2)\right] \tag{7-65}$$

$$\sigma_t = \frac{\sigma_v}{2}\left[(1 + \lambda_o)(1 + a^2) - (1 - \lambda_o)(1 + 3a^4)\right] \tag{7-66}$$

$$\tau = 0 \tag{7-67}$$

Upon loading, the hatched rock element A at $\omega = 0$ in Fig. 7-9 distance r from the center, whose side is dr, undergoes a radial deformation Δdr after excavation. Its radial strain is, by HOOKE's law:

$$\frac{\Delta dr}{dr} = \varepsilon_r = \frac{dw}{dr} = \frac{1}{E_{rock}}\left(\bar{\sigma}_{r0} - \frac{1}{m_r} \cdot \bar{\sigma}_{t0}\right) \tag{7-68}$$

where $w = $ displacement, and
$m_r = $ POISSON's number of the rock.

One should be aware that in deformation, the stress difference between primary and secondary stresses are of significance:

$$\bar{\sigma}_{r0} = \sigma_v - \sigma_{r0} \tag{7-69}$$

$$\bar{\sigma}_{t0} = \sigma_h - \sigma_{t0} \tag{7-70}$$

The zeros as indexes to the stresses indicate that these stresses pertain to $\omega = 0$, i. e., they pertain to the crown-point.

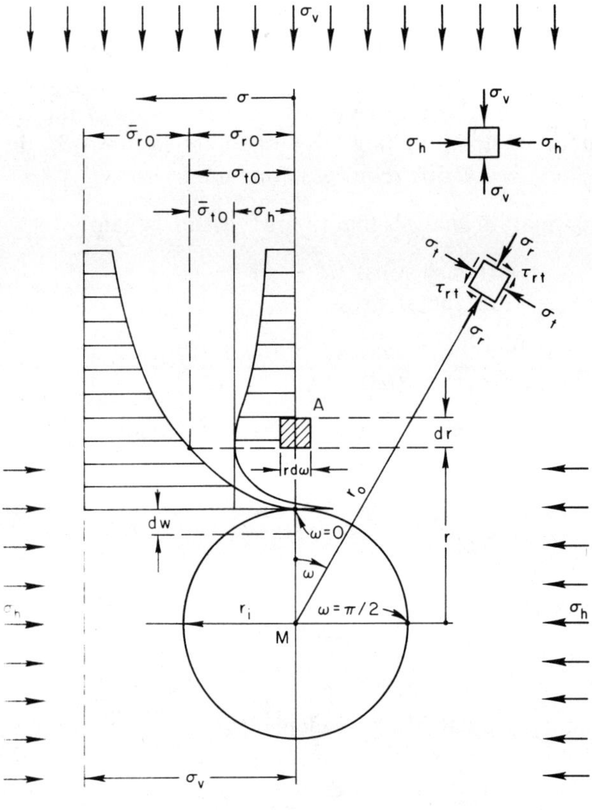

Fig. 7-9
Primary and secondary elastic principal stresses for determining elastic displacements w of the crown

Elastic Displacement w at Crest ($\omega = 0$).

By means of the stresses as given by Eqs. (7-69) and (7-70), the displacement w of the crown-point at the crest calculates as:

$$w = \int_{r_i}^{r=\infty} \Delta dr = \frac{1}{E_{\text{rock}}} \cdot \int_{r_i}^{\infty} \left[(\sigma_v - \sigma_{r0}) - \frac{1}{m_r} (\sigma_v - \sigma_{t0}) \right] dr \qquad (7\text{-}71)$$

With equations (7-65) and (7-66), and with $\lambda_o = \sigma_h/\sigma_v = 1/(m_r - 1)$, integration yields the elastic displacement w as:

$$w = \frac{\sigma_V \cdot r_i}{E_{rock}} \cdot \frac{2 \cdot m_r^2 - 3 \cdot m_r + 1}{m_r(m_r - 1)} \tag{7-72}$$

$$= \beta \cdot \frac{\sigma_V \cdot r_i}{E_r} \tag{7-72a}$$

where

$$\beta = \frac{1 - 3m_r + 2m_r^2}{m_r(m_r - 1)} \tag{7-73}$$

Elastic Displacement u at Abutment ($\omega = 90°$).

Analogically, as before, from Fig. 7-10

$$u = \int_{r_i}^{\infty} \Delta dr = \frac{1}{E_r} \int_{r_i}^{\infty} \left[(\sigma_h - \sigma_{r\,90°}) - \frac{1}{m_r}(\sigma_V - \sigma_{t\,90°}) \right] dr, \tag{7-74}$$

and

$$u = - \frac{\sigma_V \cdot r_i}{E_r} \cdot \frac{1 + m_r^2 - 4m_r}{m_r(m_r - 1)} \tag{7-75}$$

or

$$u = - \xi \cdot \frac{\sigma_V \cdot r_i}{E_r} \tag{7-75a}$$

where

$$\xi = \frac{1 + m_r^2 - 4m_r}{m_r(m_r - 1)} \tag{7-76}$$

The w- and u-deformations indicate that a circular opening in an elastic rock deforms elliptically: the vertical diameter decreases in length; the horizontal diameter increases in length (see Fig. 7-3).

7-8. Zone of Elastic Tangential Tensile Stresses

To delineate the tensile stress zone around the circular opening made in elastic rock, set the tangential stress equation, Eq. (7-42), equal to zero:

$\sigma_t = 0$. The evaluation of the $\sigma_t = 0$ equation renders the sought delineation zone or boundary line:

$$(1 + \lambda_o)(1 + a^2) - (1 - \lambda_o)(1 + 3a^4)\cos 2\omega = 0 \qquad (7\text{-}77)$$

One sees that this tensile stress is a function of POISSON's number $m = f(\lambda_o)$ only. Here $a = r_i/r$.

At the crown where $\omega = 0$,

$$a^4 - a^2 \cdot \frac{1 + \lambda_o}{3(1 - \lambda_o)} - \frac{2 \cdot \lambda_o}{3(1 - \lambda_o)} = 0 \qquad (7\text{-}78)$$

Solving for $a = r_i/r$, compute r for a given radius of circle. The difference, $r - r_i = d_{\omega = 0}$, is the thickness of the tensile zone above the crest.

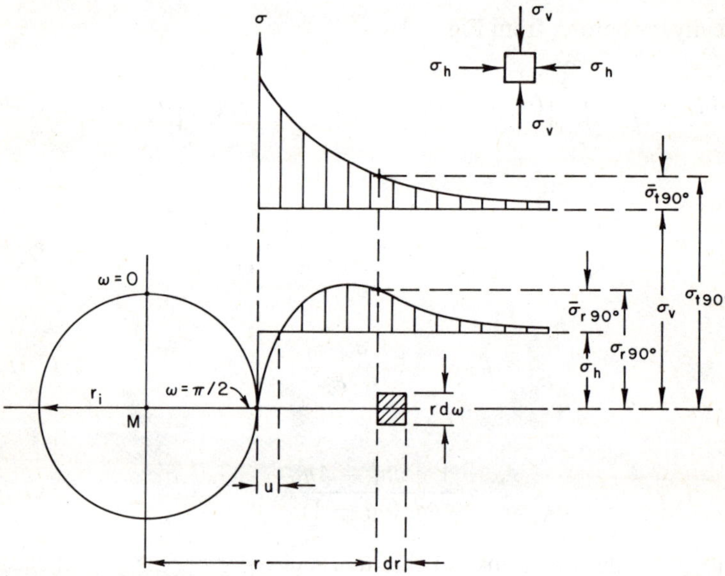

Fig. 7-10
Primary and secondary principal stresses for determining elastic displacement u of the abutment

CHAPTER 8

PLASTIC ZONES IN ROCK AROUND UNDERGROUND OPENINGS

8-1. Concept of Plastic Zone

In the discussion that follows, it is assumed that:

1. The rock material in the immediate vicinity of the tunnel is an ideally plastic material;
2. The rock lateral pressure coefficient $\lambda_o = 1.0$ and $\sigma_h = \sigma_v$, i.e., geostatic pressure distribution is assumed; and
3. Initially the rock mass was at elastic equilibrium. When the compressive strength of the rock is exceeded, around the underground opening there forms a plastic zone (Fig. 8-1).

The plastic zone is characterized by the exhaustion of the internal shear resistance of the rock up to its limit. Whether the process of yielding or plastic deformation of the rock still remains within the stage of plastic deformation movement, or whether this plastic deformation attains the point of rupture, depends upon the nature of the rock mass.

Now, for the sake of discussion, assume as valid that the plastic deformation takes place just below the limit of ultimate rupture of the rock (point C on the stress-strain diagram in Fig. 8-2).

8-2. Derivation of Plasticity Condition in Rock

Assuming a direct transition from the ideally elastic state of the rock to the ideally plastic one, let us establish equilibrium conditions that have equal validity for the elastic as well as the plastic zone of the rock mass.

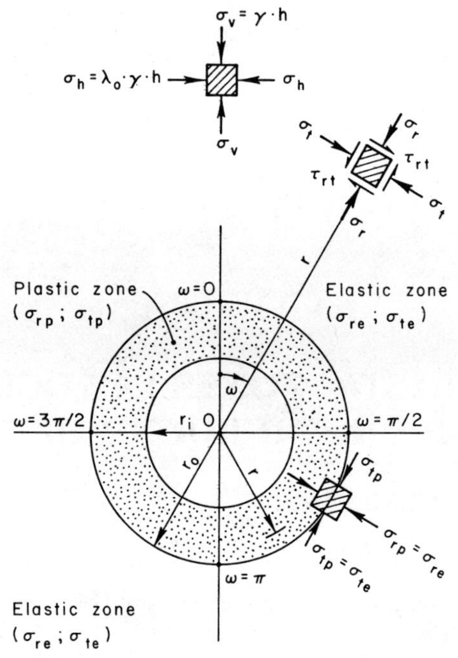

Fig. 8-1
Elastic and plastic zones

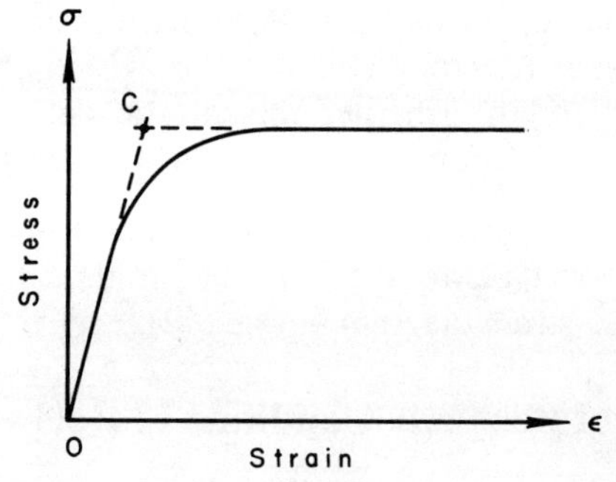

Fig. 8-2
Stress-strain diagram of an ideal plastic material

For axial, central, symmetrical stress distribution around a circular opening whose radius is r = const, there is no change in stress with ω. Therefore, in polar coordinates, the stress components, considering the effect of the weight of the rock, are independent of ω. From the theory of elasticity (11, 29), the stress components for equilibrium condition are:

$$\sigma_r = \frac{1}{r} \cdot \frac{d\Phi}{dr} \tag{8-1}$$

$$\sigma_t = \frac{d^2\Phi}{dr^2} \tag{8-2}$$

$$\tau = 0 \tag{8-3}$$

In these equations, the symbol Φ represents AIRY's stress function. However, in order to describe the stress condition in rock, equilibrium equations (8-1) and (8-2) alone are insufficient, although they are necessary. The use of the AIRY's stress function opens the way to determine the stresses if one more conditions is given. Recall that for the elastic zone, HOOKE's law of proportionality between stress and strain (*"ut tensio, sic vis"*) is valid. For

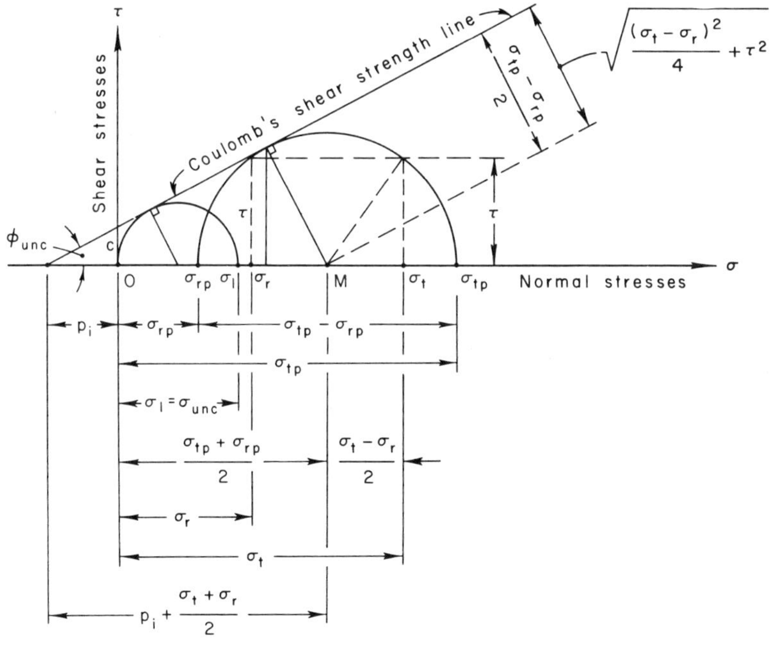

Fig. 8-3
Plasticity condition in COULOMB-MOHR presentation

the plastic zone, the condition of the internal resistance of the rock against shear at the ultimate limit (MOHR's theory of failure) is utilized. This is the condition when COULOMB's shear strength line tangents MOHR's stress circles (Fig. 8-3). This line depends upon the uniaxial compressive strength σ_{unc} of the rock mass and the test parameter φ_{unc}, popularly known as the angle of internal friction. The radial and tangential stresses, σ_{rp} and σ_{tp}, respectively, in the plastic zone can be determined by means of MOHR's theory of failure. Here the index p refers to stresses in the plastic condition.

The plasticity condition can now be written from MOHR's stress circle diagram, as in Fig. 8-3 (11, 16):

$$\sin \varphi_{unc} = \frac{\sigma_{tp} - \sigma_{rp}}{2p_i + \sigma_{rp} + \sigma_{tp}} \tag{8-4}$$

where φ_{unc} = stress parameter (angle of internal friction as obtained in the unconfined compression test (unc),

$p_i = c \cdot \cot \varphi_{unc}$ = initial stress in rock, and

c = cohesion of rock.

On the other hand:

$$\sin \varphi_{unc} = \frac{\sigma_{unc}}{2p_i + \sigma_{unc}} \tag{8-5}$$

and

$$p_i = c \cdot \cot \varphi_{unc} = \frac{\sigma_{unc}}{2} \cdot \frac{1 - \sin \varphi_{unc}}{\sin \varphi_{unc}} \tag{8-6}$$

where σ_{unc} = unconfined compressive strength of rock (see Fig. 8-3).

Substitution of (8-6) into (8-4) renders:

$$\sigma_{tp} - \frac{1 + \sin \varphi_{unc}}{1 - \sin \varphi_{unc}} \cdot \sigma_{rp} - \sigma_{unc} = 0 \tag{8-7}$$

With

$$(1 + \sin \varphi_{unc})/(1 - \sin \varphi_{unc}) = \lambda_{rp} = \zeta \tag{8-8}$$

where $\lambda_{rp} = \zeta$ is the coefficient at σ_{rp} in Eq. (8-7), the plasticity condition takes the form:

$$\sigma_{tp} - \zeta \cdot \sigma_{rp} - \sigma_{unc} = 0 \tag{8-9}$$

Introducing the stresses from the equilibrium condition — Eqs. (8-1) and (8-2) — into the plasticity condition — Eq. (8-9) — obtain a differential equation of the AIRY's stress function Φ:

$$\frac{d^2\Phi}{dr^2} - \zeta \cdot \frac{1}{r} \cdot \frac{d\Phi}{dr} - \sigma_{unc} = 0 \tag{8-10}$$

As shown by KASTNER (16), integration of Eq. (8-10) renders the sought AIRY's stress function Φ:

$$\Phi = C_1 \cdot \frac{r^{\zeta+1}}{\zeta+1} - \frac{\sigma_{unc}}{\zeta-1} \cdot \frac{r^2}{2} + C_2 \tag{8-11}$$

$$\frac{d\Phi}{dr} = C_1 \cdot r^{\zeta} - \frac{\sigma_{unc}}{\zeta-1} \cdot r \tag{8-12}$$

$$\frac{d^2\Phi}{dr^2} = C_1 \cdot r^{\zeta-1} - \frac{\sigma_{unc}}{\zeta-1} \tag{8-13}$$

The boundary conditions are:

at $r = r_i$, $\sigma_{rp} = 0$, yielding the value of the integration constant C_1 as:

$$C_1 = \frac{1}{r_i^{\zeta-1}} \cdot \frac{\sigma_{unc}}{\zeta-1} \tag{8-14}$$

The integration constant C_2 does not enter into the equation of the plasticity condition.

With C_1, C_2, Φ, $d\Phi/dr$, and $d^2\Phi/dr^2$ now known, the stress components of the plastic zone are:

$$\sigma_{rp} = \frac{\sigma_{unc}}{\zeta-1} \cdot \left[\left(\frac{r}{r_i}\right)^{\zeta-1} - 1\right] \tag{8-15}$$

$$\sigma_{tp} = \frac{\sigma_{unc}}{\zeta-1} \cdot \left[\zeta \cdot \left(\frac{r}{r_i}\right)^{\zeta-1} - 1\right] \tag{8-16}$$

$$\tau_p = 0 \tag{8-17}$$

8-3. Extent of Plastic Zone

The plastic zone is delineated from the elastic zone by a circle whose radius is r_o (Fig. 8-1).

By Eqs. (7-28), (7-56), and (7-57), the elastic stresses in the elastic zone induced by the all-sided, equal, primary, compressive stress $\sigma = \sigma_v = \sigma_h$ are:

$$\sigma_{rel} = \sigma\,(1 - \alpha^2) \tag{8-18}$$

$$\sigma_{te1} = \sigma(1 + \alpha^2) \tag{8-19}$$

$$\tau_{e1} = 0 \tag{8-20}$$

Herein, the index e pertains to the elastic stress; the index '1' refers to the primary (initial) stress condition, and $\alpha = r_o/r$ [Eq. (7-28)].

Besides primary stresses in the initial stress condition, there also prevails in the elastic zone a yet unknown axial-symmetrical radial stress σ_{r_o}. Thus, along the circular boundary of radius r_o between the elastic and plastic zones, these additional stresses are:

$$\sigma_{re2} = \sigma_{r_o} \cdot \frac{r_o^2}{r^2} \tag{8-21}$$

$$\sigma_{te2} = -\sigma_{r_o} \frac{r_o^2}{r^2} \tag{8-22}$$

$$\tau_{e2} = 0 \tag{8-23}$$

Summation of stresses (8-18) and (8-21), (8-19) and (8-22), and (8-20) and (8-23) renders the total elastic stresses on the elastic-plastic boundary:

$$\sigma_{re} = \sigma_{re1} + \sigma_{re2} = \sigma\left(1 - \frac{r_o^2}{r^2}\right) + \sigma_{r_o} \cdot \frac{r_o^2}{r^2} \tag{8-24}$$

$$\sigma_{te} = \sigma_{te1} + \sigma_{re2} = \sigma\left(1 + \frac{r_o^2}{r^2}\right) - \sigma_{r_o} \cdot \frac{r_o^2}{r^2} \tag{8-25}$$

$$\tau_e = \tau_{e1} + \tau_{e2} = 0 \tag{8-26}$$

At the elastic-plastic boundary zone $\left(r = r_o \text{ or } \frac{r_o}{r} = 1\right)$ the prevailing stresses should be equal, such as:

$$\sigma_{r_o} = \sigma_{rp} = \sigma_{re} \tag{8-27}$$

and

$$\sigma_{tp} = \sigma_{te} \tag{8-28}$$

Introducing Eq. (8-27) into Eq. (8-15), obtain

$$\frac{\sigma_{unc}}{\zeta - 1} \cdot \left[\left(\frac{r_o}{r_i}\right)^{\zeta-1} - 1\right] = \sigma_{r_o} \tag{8-29}$$

$$\frac{\sigma_{unc}}{\zeta - 1} \cdot \left[\zeta \cdot \left(\frac{r_o}{r_i}\right)^{\zeta-1} - 1\right] = 2\sigma - \sigma_{r_o} \tag{8-30}$$

Herein:

$$\zeta = \frac{1 + \sin\varphi_{unc}}{1 - \sin\varphi_{unc}} \tag{8-8}$$

Upon adding Eqs. (8-29) and (8-30), the σ_{r_o} stress cancels out, resulting in

$$\left(\frac{r_o}{r_i}\right)^{\zeta-1} \cdot (\zeta + 1) = \frac{2\sigma(\zeta - 1)}{\sigma_{unc}} + 2 \tag{8-31}$$

From this equation, the boundary (radius r_o) of the extent of the plastic zone calculates as:

$$r_o = r_i \left[\frac{2}{\zeta + 1} \cdot \frac{\sigma(\zeta - 1) \cdot \sigma_{unc}}{\sigma_{unc}}\right]^{\frac{1}{\zeta-1}} \tag{8-32}$$

This equation shows that:

$$r_o = f(r_i;\ \lambda_o = 1;\ \sigma = \sigma_v = \sigma_h = \gamma \cdot h;\ \sigma_{unc};\ \varphi_{unc}) \tag{8-33}$$

The thickness of the plastic zone is thus $\Delta r = r_o - r_i$.

Summary of stresses for $\lambda_o = 1$

$$\sigma_{rel} = \sigma(1 - \alpha^2) = \sigma\left(1 - \frac{r_o^2}{r^2}\right) \tag{8-18}$$

$$\sigma_{tel} = \sigma(1 + \alpha^2) = \sigma\left(1 + \frac{r_o^2}{r^2}\right) \tag{8-19}$$

$$\tau_{el} = 0 \tag{8-20}$$

$$\sigma_{re2} = \sigma_{r_o} \cdot \frac{r_o^2}{r^2} \tag{8-21}$$

$$\sigma_{te2} = -\sigma_{r_o} \cdot \frac{r_o^2}{r^2} \tag{8-22}$$

$$\tau_{e2} = 0 \tag{8-23}$$

At the plastic-elastic boundary ($r = r_o$)

$$\sigma_{r_o} = \sigma_{re} = \sigma_{rp};\ \sigma_{tp} = \sigma_{te} \tag{8-28}$$

$$\sigma_{r_o} = \sigma_{rp} = \frac{\sigma_{unc}}{\zeta - 1} \cdot \left[\left(\frac{r_o}{r_i}\right)^{\zeta-1} - 1\right] \tag{8-29}$$

$$2\sigma - \sigma_{r_o} = \frac{\sigma_{unc}}{\zeta - 1}\left[\zeta\left(\frac{r_o}{r_i}\right)^{\zeta-1} - 1\right] \tag{8-30}$$

Stresses in the plastic zone

$$\sigma_{rp} = \frac{\sigma_{unc}}{\zeta - 1}\left[\left(\frac{r}{r_o}\right)^{\zeta-1} - 1\right] \tag{8-34}$$

$$\sigma_{tp} = \frac{\sigma_{unc}}{\zeta - 1}\left[\zeta\left(\frac{r}{r_o}\right)^{\zeta-1} - 1\right] \tag{8-35}$$

$$\tau_p = 0 \tag{8-36}$$

8-4. Discussion on Stresses in Elastic and Plastic Zones

For the sake of the discussion that follows, refer to Fig. 8-4 after KASTNER (16). In this figure are shown radial and tangential stress distributions in elastic and plastic zones in rock around a horizontal, circular underground opening. This figure pertains to the following example.

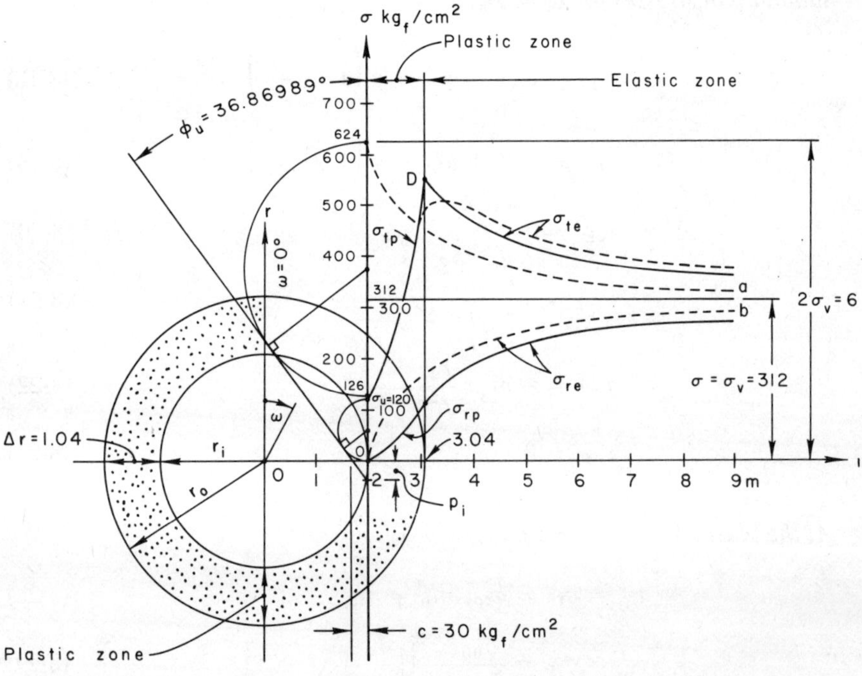

Fig. 8-4
Radial and tangential stresses around a horizontal, circular underground opening in elastic and plastic zones of sound rock

Assume a sound rock whose unit weight is $\gamma = 2.60\ t_f/m^3$. Its unconfined compressive strength is $\sigma_1 = \sigma_u = 120\ kg_f/cm^2$, and the test parameter $\varphi_u = 36.86989°$, resulting in a shear strength at zero pressure of $s = c = 30\ kg_f/cm^2$, and in an initial stress of $p_i = c \cdot \cot \varphi_u = (30)\,(1.33333) = 40\ [kg_f/cm^2]$. These strength data are shown on the Coulomb-Mohr strength diagram in Fig. 8-4.

The thickness of the overburden rock surcharge is given as $h = 1200$ m. Thus, the all-sided, uniform, overburden pressure for $\lambda_o = 1.0$ is:

$$\sigma = \sigma_v = \sigma_h = \gamma \cdot h = (2.6)\,(1200) = 3120\ [t_f/m^2]$$

$$= 312\ [kg_f/cm^2]$$

shown in Fig. 8-4.

The inside radius of the opening is given as $r_i = 2.00$ m. By Eq. (8-7), for $\zeta = (1 + \sin 36.86989°)/(1 - \sin 36.86989°) = 4.0$, the outside radius r_o of the plastic zone calculates as $r_o = 304$ cm $= 3.04$ m. Thus, the thickness of the annular ring of the plastic zone is $\Delta r = r_o - r_i = 3.04 - 2.00 = 1.04$ [m].

For the plastic zone not to occur, $r_o = r_i$ (for $\lambda_o = 1$ and $a = a = r_o/r$ 1.0), the stresses at the contour for $\omega = 90°$ are, by Eqs. (7-56), (7-57), and (7-58):

$$\sigma_r = 0$$
$$\sigma_t = 2\,\sigma_v = 2\sigma = (2)\,(312) = 624\ [kg_f/cm^2] = \sigma_{te} = \sigma_{tp}$$
$$\tau = 0$$

The stresses for the plastic zone and for the elastic zone are computed by Eqs. (8-15) and (8-18), respectively.

By Eq. (8-15), the radial plastic stress σ_{rp} corresponding to the tangential plastic stress σ_{tp} calculates as 126 kg_f/cm^2.

For the conditions as indicated above ($\lambda_o = 1.0$, $a = 1.0$, and $\omega = 90°$), the distribution of tangential and radial stresses around a horizontal, circular opening in plastic and elastic zones at all-sided primary, uniform pressure is sketched in Fig. 8-4.

In Fig. 8-4, the dashed curve marked as "*a*" pertains to tangential stress distribution in a sound, elastic rock. Likewise, the dashed curve marked as "*b*" pertains to radial stress distribution in a sound, elastic rock.

The studies on plastic deformations of rock justify the following observations:

1. The problem of rock pressure on an underground opening represents a particular phase in the discipline of rock mechanics.

2. The determination of the magnitude of rock pressure is one of the most intricate problem in rock mechanics.
3. The kind of rock stress depends upon many factors, especially on the quality of the rock, and on the depth of the underground opening below ground surface.
4. In the aforegoing discussion, only the strength properties of the rock material are considered. They are unaffected by that rock pressure for which a geostatic pressure distribution ($\sigma = \sigma_v = \sigma_h = \gamma h$) is assumed.
5. Because of excavating of an underground opening in a rock massif, secondary stresses in rock are brought about, resulting in stress redistribution in the rock around the opening, and plastic deformation in the rock sets in. A plastic zone develops in the rock between the contour of the opening and the elastic zone. The extent of the plastic zone depends to a considerable degree on the ratio ζ of vertical to lateral pressures.
6. The boundary between the plastic and elastic zones is assumed to be a circular zone, labeled by r_0 (outer radius of the plastic zone).
7. At the plastic and elastic zone boundary ($r = r_0$), the elastic and plastic stresses as computed by the theories of elasticity and plasticity must be equal.
8. Generally, near the contour of the opening, there prevail high stress peaks in the rock. Because of plastic deformations and stress redistribution, the high contour stresses have a tendency to decrease gradually.
9. Also, because of plastic deformations and consequent stress redistribution, wider rock zones around the opening become stressed.
10. From Fig. 8-4, it can be observed that at the plastic-elastic boundary ($r = r_0$) the tangential stress diagram has a discontinuity (point D), where it exhibits a peak value. This peak value is in excess of the strength of the rock, leading to a plastic stress condition. Actually, between the plastic and active zones, in the vicinity of the discontinuity at D, especially for the tangential stresses, there will be most likely a transition zone of rounding up these stresses *(natura non facit saltus)* thus shifting the tangential stress toward the interior of the elastic rock mass.
11. After the peak value, the magnitude of the tangential stress decreases in a certain proportion with distance away from the contour of the opening.

12. Depending upon the kind and quality of the rock, the tangential stress peak may attain a multiple of the magnitude of the average vertical stress. At some depth, such a tangential peak stress may even exceed the strength of the rock massif.
13. Within the plastic zone, the magnitude of the radial stress increases with distance toward the interior of the elastic rock massif. This may be explained as follows. The particles of the plastic rock material are subjected to the normal tangential stress as a normal force. Because of friction between the particles, these particles become more and more capable of resisting radial shear stresses.
14. It should not be forgotten that the assumption $\lambda_0 = 1.0$ pertains to an ideal condition only. Nonetheless, the application of $\lambda_0 = 1.0$ or nearly 1.0: (a) affords a rigorous theoretical plasticity analysis, rendering a good idea about the pressure condition within the rock massif, and (b) may be considered to be a good basis for designing the size of a tunnel lining.

8-5. Zone of Disturbance

In reality, upon making an underground opening in rock (say, by means of explosives), unfavorable stress conditions in the rock are introduced in the immediate proximity around the opening. Upon blasting out an opening, the rock here becomes disintegrated. As a consequence, a loose zone, or zone of disturbance between the contour of the opening and the plastic zone, is brought about (Fig. 8-5). Because of the disintegrated condition of the

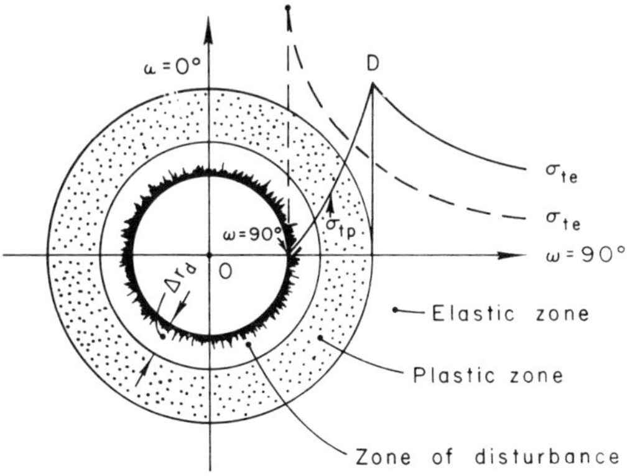

Fig. 8-5
Course of stress redistribution in rock brought about by a loose zone of rock disturbance

rock in this zone of disturbance, the rock here is not in a plastic state. It does not have any unconfined compressive strength, nor can it take and carry any stress.

Because of the loose zone, the stresses shift and redistribute farther away from the disturbed, loose zone into the rock medium. The value of the tangential stresses at the contour reduces to zero.

The thickness of the zone of blasting disturbance is designated as Δr_d. In order for this loose zone to exist, the opening must be braced or strutted.

Without support, the zone of disturbance may exist only by arching of the rock.

In Fig. (8-5), the stress distribution diagram shows an increase in the tangential stress from zero at the contour up to a peak (marked by D). This peak is attained at the boundary between the plastic and elastic zones. Farther away from the plastic-elastic boundary in the direction inward in the rock, the rock is affected only by elastic deformations. Thus, the rest of the course of the stress curve is similar to that shown in Fig. (8-4) for the case of plastic-elastic zones with no loose zone of disturbance.

The shift of the stresses displaces the centroid of the area of the stress diagram toward the interior of the rock, viz., away from the opening. This would mean nothing else than the natural self-aid of the rock material, in that the more remote ranges of the rock massif are being, so to say, mobilized to help to carry the stresses.

8-6. General Notes about Slip Lines

The phenomenon of slip, or displacement by translation of soil or rock particles, consists of a sliding (translocation) of particles along planes or surfaces of weakness relative to one another. This sliding of particles occurs in numerous planes, viz., surfaces forming a certain geometric slip line pattern. This type of deformation is often evidenced by numerous, remarkable parallel and mutually interesting markings on the mantle surface of soil specimens subjected to compression tests (Fig. 8-6). See also Reference 11.

Such markings or flow figures showing geometric patterns of families of mutually intersecting slip lines can also be observed on the cylindrical mantle surface of radially deformed rock specimens whenever the yield has reached in them under a radially symmetrical state of stress.

By the expression "geometric pattern of slip lines is the state of plane plastic strain" is understood two systems of plane curves in which the cylindrical surfaces of slip normal to the x,y plane intersect the latter plane.

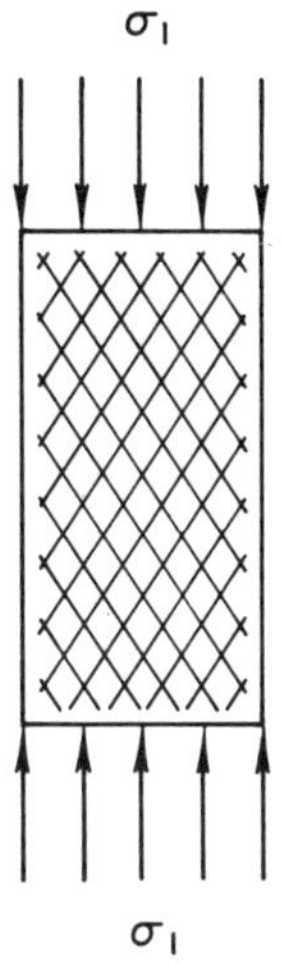

Fig. 8-6
Mutually intersecting markings on mantle surface of a soil specimen in the state of plastic equilibrium

For a two-dimensional case, a curve whose tangent at every point is one of the shear directions at that point on the curve is variously called in the literature a slip line, or shear line, or line of rupture, or shear stress trajectory. Slip lines are the most useful means for representing the stress field.

In tackling rock mechanics problems as well as in dealing with pertinent problems in many other geotechnical engineering disciplines, in the evaluation of the shear stress conditions in or shear strength of the material, the engineer's interest centers about the determination of the shear stress in the rupture surface, and on the delineation of the shear stress lines and the rupture surface as well.

These shear stress lines are the subsequent slip lines, often called the lines of rupture, or rupture surface curves.

As a recourse from the theory of plasticity, it should be remembered that, in general, at limit equilibrium, there is a set of two impending rupture surface curves, or special slip lines, through every point in the material.

Knowledge of slip lines can be used to advantage for studying stress conditions and rupture surfaces in soil and rock materials brought about by structural loads.

Proceeding in the rock and soil mass from point to point, the direction of action of the principal stresses $\sigma_1 = \sigma_{tp}$ and $\sigma_3 = \sigma_{rp}$ and the course of slip-line curves in the material can be traced. Such a tracing results in a network or family of curves such as the slip lines or shear stress trajectories. Tangents drawn to the shear stress curves would then show the direction of action of the major principal stress σ_1 (for example, at the given point of tangency). The minor principal stress σ_3 then acts at this point in the direction normal to the tangent — that is, normal to σ_1.

8-7. Form of Slip Lines

For a rotary-symmetrical plastic zone such as one around a horizontal, circular underground opening, the slip line form is derived mathematically as set forth.

At primary, all-sided uniform pressure ($\lambda_0 = 1.0$), the principal stresses, σ_{rp} and σ_{tp}, in the plastic zone are given by Eqs. (8-15) and (8-16) as:

$$\sigma_{rp} = \frac{\sigma_{unc}}{\zeta - 1} \cdot \left[\left(\frac{r}{r_i} \right)^{\zeta - 1} - 1 \right] \tag{8-15}$$

$$\sigma_{tp} = \frac{\sigma_{unc}}{\zeta - 1} \cdot \left[\zeta \cdot \left(\frac{r}{r_i} \right)^{\zeta - 1} - 1 \right] \tag{8-16}$$

The shear stress in this zone is zero: $\tau_p = 0$.

From Mohr's stress diagram in Fig. (8-7b), Coulomb's shear strength line OT — a tangent to the stress circle — as a boundary line represents the plastic flow condition, viz., failure, of the rock material. The directions of rupture surfaces 1-1 and 2-2 form angles of rupture ($+\,a$) and ($-\,a$) with the stress axis (σ), such that:

$$\tan \alpha = \frac{\left(\frac{\sigma_{tp} - \sigma_{rp}}{2} \right) \cdot \cos \varphi}{\left(\frac{\sigma_{tp} - \sigma_{rp}}{2} \right) (1 - \sin \varphi)} = \frac{\cos \varphi}{1 - \sin \varphi} = f(\varphi) = \text{const} \tag{8-37}$$

where φ = angle of internal friction of the in-situ rock massif. The principal stresses, σ_{tp} and σ_{rp}, are to be computed by Eqs. (8-15) and (8-16).

Equation (8-37) indicates that the form of the rupture surfaces is equilateral logarithmic spirals. Their assymptote is the pole (P) of the polar coordinate system, viz., the center M of the circular opening (Figs. 8-7a and

a) Slip line tangents 1—1 and 2—2 at point S in the plastic zone

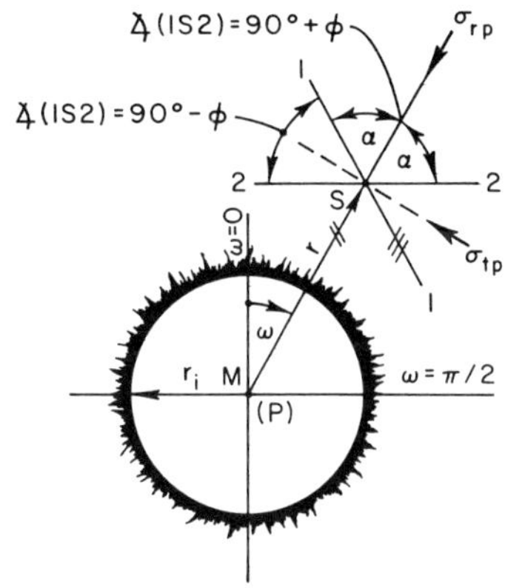

b) MOHR's stress circle for plastic equilibrium

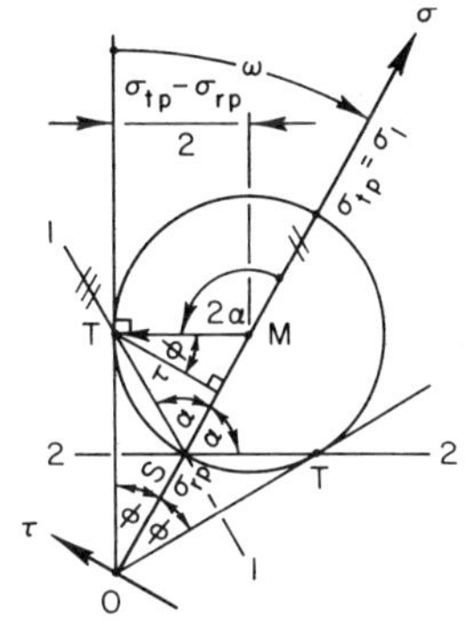

c) System of equilateral logarithmic spirals

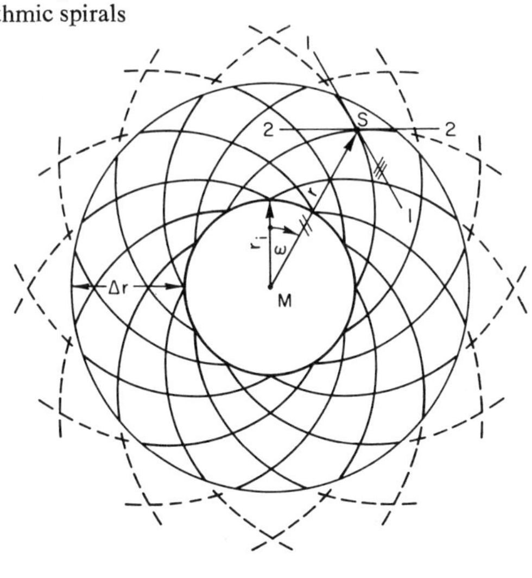

Δr = thickness of plastic zone

Fig. 8-7
Slip lines in the plastic zone

c). Therefore, the equation of this logarithmically spiraled rupture surface curve can be written at once as

$$r = r_i \cdot e^{\left(\frac{1}{\tan \alpha}\right) \cdot \omega} \tag{8-38}$$

$$= r_i \cdot e^{\left(\frac{1 - \sin \varphi}{\cos \varphi}\right) \cdot \omega}$$

or

$$r = r_i \cdot e^{\pm \omega \cdot \tan \varphi} \tag{8-39}$$

(see Ref. 32),
where

$$\tan \varphi = \frac{1}{\tan \alpha} = \frac{1 - \sin \varphi}{\cos \varphi}, \tag{8-40}$$

r and ω = polar coordinates, and

r_i = initial radius vector = inner radius of the opening.

Thus, if r_i and φ are known, the course of the slip lines, and thus their pattern in the plastic zones around a circular opening at primary, all-sided uniform pressure, can be traced by Eq. (8-39). The form of the logarithmic spiral is thus governed by the value of tan φ.

For the construction of the spiral, the magnitudes of the r-vectors are to be computed for each of the selected amplitudes ω with various lengths of radii starting with $r = r_i$ at the contour and ending up with $r = r_o$ at the plastic-elastic boundary.

The plotting of the logarithmically spiraled slip lines is greatly facilitated by means of the author's physical, dimensionless logarithmic spiral tables (10).

Figure (8-7c) shows a system of equilateral, logarithmically spiraled slip lines or shear stress trajectories in a plastic zone around a horizontal, circular underground opening for $\varphi = 30°$, and $\frac{\sigma_{tp} - \sigma_{rp}}{2} = 200$ kg$_f$/cm^2, computed for every $\omega = 30°$.

8-8. Summary about Slip Lines

1. The slip lines are equilateral, logarithmic spirals.

2. The form of the logarithmic spiral depends upon the parameter φ, viz., value of tan φ.

3. The parameter r_i (initial radius) has no effect on the form of the spiral.

4. The family of slip lines intersects mutually.

5. No slip line in a homogeneous, isotropic rock mass has the advantage over the other in the same system of slip-line pattern.

6. The slip lines, viz., slip surfaces, terminate at the boundary between the plastic and elastic zones where $r = r_0$. Therefore, this zone boundary is not a slip surface.

7. The theory of plasticity has shown that the slip lines may be used with great advantage to represent the nature, condition, direction, and comparative amount of the shear stresses.

8. If the assumption of isotropy is not met, then preferred slip surfaces will prevail, dictated by the anisotropy.

CHAPTER 9

STRESSES IN ELASTIC ROCK AROUND VERTICAL SHAFTS

9-1. Primary Stress Conditions in the Elastic Zone

Prior to making a shaft, or drilling a borehole in a rock mass in situ, i.e., prior to disturbance, it is visualized that the rock mass is subjected to the effects of some of the rock-forming forces as well as to the force of gravity.

The rock-forming forces are difficult to comprehend. One generally assumes that the effects of the rock-forming forces (tectonic forces, for example) have already died or faded out, and that only the effect of the gravity force is usually considered in rock stress analysis of shafts, boreholes, galleries and tunnels made in rock.

In order to simplify calculations, one further assumes a homogeneous geologic material such as soil or rock the unit weight of which is γ. On a horizontal plane at depth z below the ground surface, there acts, then, a vertical compressive stress, $\sigma_v = \sigma_z = \gamma \cdot z$, known in rock mechanics as characterizing the primary stress field in the rock mass. Furthermore, as a consequence of the above assumption, it is assumed that at a given depth $z = h$, there acts a uniformly distributed lateral pressure from the rock mass:

$$\sigma_h = \sigma_x = \gamma \cdot z \tag{9-1}$$

around the outer circular contour of the shaft.

If there is a possibility for the rock mass to expand laterally then there occurs a corresponding lateral strain ε_x. However, the subsurface mass of rock, assumed in the theory of elasticity to be an infinitely extending hemi-spatial medium, cannot expand. Therefore, the lateral strain ε_x in HOOKE's stress-strain relationship is equal to zero, i.e.:

$$\varepsilon_x = \frac{1}{E} \cdot \left[\sigma_x - \frac{1}{m}(\sigma_y + \sigma_z)\right] = 0 \tag{9-2}$$

where E = modulus of elasticity of rock, viz., soil;
m = POISSON's number; and
σ_x and σ_y = lateral stresses (σ_h) at depth z below ground surface.

The two lateral stresses, $\sigma_x = \sigma_y = \sigma_h$, are considered to be of equal magnitude. Therefore, by Eq. (9-2):

$$\sigma_x = \sigma_y = \sigma_h = \frac{\sigma_z}{m-1} = \frac{\gamma \cdot z}{m-1} \tag{9-3}$$

Thus, assuming that the rock-forming forces are equal to zero, Eqs. (9-1) and (9-3) represent the undisturbed state of stress condition in a rock or soil formation at depth z prior to sinking a shaft or making a boring in the ground.

9-2. Secondary Elastic Stress Conditions

Upon sinking a shaft or drilling a borehole, the undisturbed stress condition in the rock or soil becomes considerably disturbed and altered. The new, or secondary elastic stress condition, for a central-symmetrical system, for any arbitrary depth z is calculated by means of AIRY's stress function, and is characterized in polar coordinates as set forth.

Radial stress:

$$\sigma_r = \frac{1}{2}\left(A - \frac{B}{r^2}\right) \tag{9-4}$$

Tangential stress:

$$\sigma_t = \frac{1}{2}\left(A + \frac{B}{r^2}\right) \tag{9-5}$$

Shear stress:

$$\tau = 0. \tag{9-6}$$

Here r is a variable radius (Fig. 9-1).

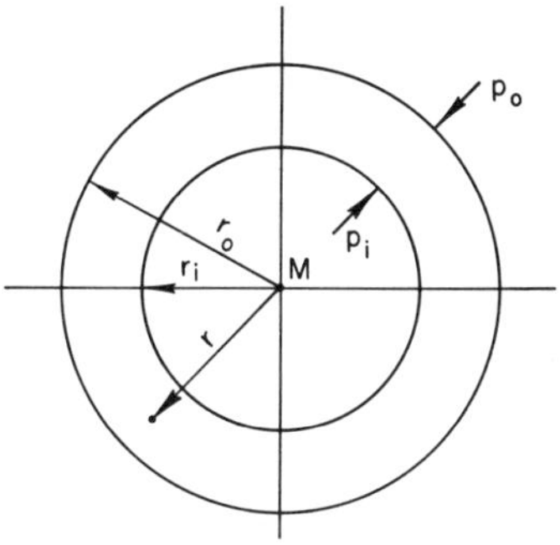

Fig. 9-1
Horizontal section of vertical, circular opening in the ground at depth z

The constants A and B are determined by means of the following boundary conditions:

1. When $r \rightarrow \infty$, then:

$$\sigma_r = \sigma_t = \sigma_x = \sigma_y = \sigma_h = \frac{\gamma \cdot z}{m - 1} \tag{9-7}$$

2. When $r \rightarrow r_o$, then $\sigma_r = 0$

$$\therefore \quad \sigma_r = \frac{A}{2} = \frac{\gamma z}{m - 1} = \sigma_t \tag{9-8}$$

$$\therefore \quad A = \frac{2 \cdot \gamma \cdot z}{m - 1} \tag{9-9}$$

$$\therefore \quad \sigma_r = 0 = \frac{1}{2}\left(\frac{2 \cdot \gamma \cdot z}{m - 1} - \frac{B}{r_o^2}\right) \tag{9-10}$$

$$\therefore \quad B = \left(\frac{2\gamma z}{m - 1}\right) \cdot r_o^2 \tag{9-11}$$

Hence:

$$\sigma_r = \frac{1}{2}\left(\frac{2\gamma z}{m - 1} - \frac{2\gamma z}{m - 1} \cdot \frac{r_o^2}{r^2}\right) \tag{9-12a}$$

or

$$\sigma_r = \frac{\gamma z}{m - 1}\left(1 - \frac{r_o^2}{r^2}\right) = \sigma_h\left(1 - \frac{r_o^2}{r^2}\right) \tag{9-12}$$

$$\sigma_t = \frac{\gamma z}{m - 1}\left(1 + \frac{r_o^2}{r^2}\right) = \sigma_h\left(1 + \frac{r_o^2}{r^2}\right) \tag{9-13}$$

Equations (9-12) and (9-13) represent the new, or the secondary stress condition in a soil or a rock mass at a particular depth z upon making a vertical, circular underground opening.

Because the shear stresses are equal to zero for reasons of central symmetry, the tangential stress σ_t is the major principal stress, and the radial stress σ_r is the minor principal stress. The vertical stress $\sigma_z = \sigma_v = \gamma \cdot z$ is an intermediate principal stress.

The theoretical secondary elastic stress conditions σ_r and σ_t in an elastic rock or soil medium beyond the contour of a vertical shaft or a borehole are shown in Fig. 9-2. These stress conditions are good as long as the limit of elasticity of the material is not exceeded. The minute the limit of the elastic deformations of the rock is exceeded, the foregoing assumptions made for these elastic stress condition calculations are no more valid. Therefore, in such a case, our knowledge about the prevailing real stress conditions and behavior of the rock or soil is confined to one's qualitative reasoning.

9-3. Discussion

1. The assumption of homogeneity of rock does not correspond to the reality. The rock is usually fractured and fissured, and the bedding planes of the rock formations are not always horizontal, nor parallel. Recall rock defects.

2. The shaft is usually blasted out by means of explosives. Hence the inside wall around the contour (circumference) of the shaft is irregularly formed, the rock is broken, and fissures may be brought about. They may propagate far out radially as well as deep in depth of the rock.

3. Equations (9-1), (9-12) and (9-13) show that all stress components, σ_r, σ_t and $\sigma_z = \sigma_v$, are compressive stresses.

4. The stress components σ_r and σ_t do not depend on the diameter of the shaft or the borehole, but on the (r_i/r)-ratio.

5. With increase in distance r from the vertical axis of the shaft, the radial stress σ_r increases, and the tangential stress σ_t decreases (Fig. 9-2).

6. At the contour, $r = r_i = r_o$, therefore, by Eqs. (9-12) and (9-13):

$$\sigma_r = 0$$

and

$$\sigma_t = \frac{2\gamma z}{m-1} = 2\sigma_h = \frac{2\sigma_v}{m-1} = 2 \cdot \sigma_x = \text{const} \tag{9-14}$$

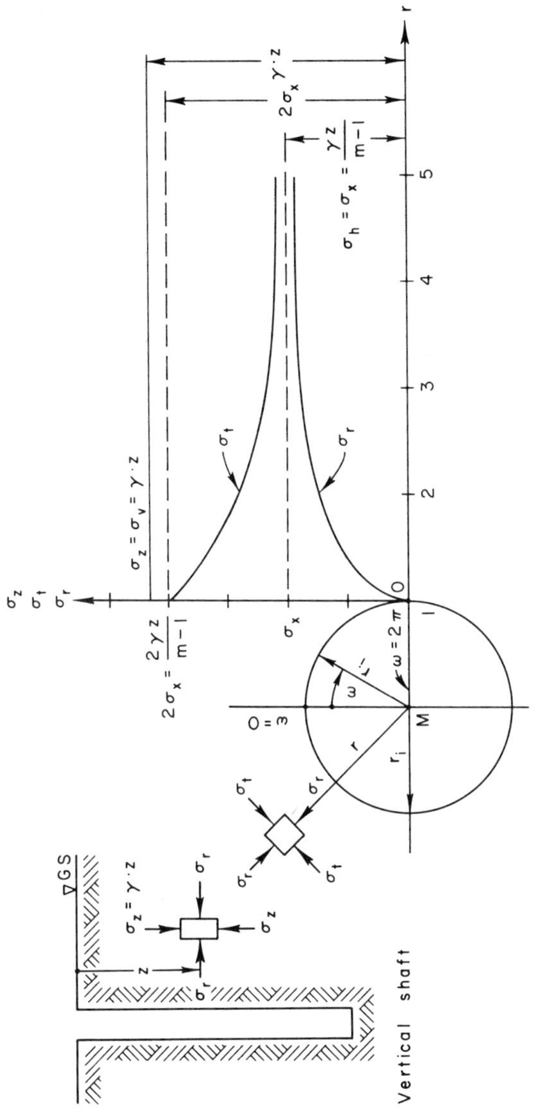

Fig. 9-2
Elastic stress conditions in an elastic medium of rock around a vertical, circular shaft at depth *z*

Hence, at the contour of the shaft, the radial and tangential stresses σ_r and σ_t, respectively, do not depend on the *(r_i/r)*-ratio of the shaft, but for a definite depth *z* below ground surface, and for a given POISSON's number *m* on hand, are constant quantities:

$$\sigma_r = 0; \quad \sigma_t = (2 \cdot \sigma_v)/(m - 1) = 2\sigma_h; \quad \sigma_v = \sigma_z = \gamma \cdot z. \tag{9-15}$$

Notice that in a sound rock, the tangential stress σ_t has the double value of the horizontal stress $\sigma_h = \sigma_x$.

7. At the contour, the increase and decrease in σ_r and σ_t, respectively, is such that the sum of these two stresses at each point on the contour for a certain depth *z* is equal to $2\sigma_h = 2\sigma_x$, i.e.,

$$\sigma_t + \sigma_r = \frac{2\gamma z}{m - 1} = 2\sigma_x \tag{9-16}$$

8. Because the displacements of the rock are directed radially toward the shaft, and because simultaneously no relative displacement between the individual particles of soil or rock mass can take place, physically the shear stress is equal to zero. Therefore, the radial and tangential directions represent the direction of action of principal stresses σ_r and σ_t.

9. The directions of sliding planes intersect the lines of action of principal stresses, viz., principal planes at a constant angle.

10. The shear stress is the same along the contour of the shaft. In general, the maximum shear stress τ_{max} is equal to one-half of the difference between the major and minor principal stress. In this particular case:

$$\tau = \frac{\sigma_1 - \sigma_3}{2} = \frac{\sigma_t - \sigma_r}{2} \tag{9-17a}$$

$$= \frac{\gamma z}{2\,(m - 1)} \left[\left(1 + \frac{r_i^2}{r^2}\right) - \left(1 - \frac{r_i^2}{r^2}\right) \right]$$

$$= \frac{\gamma z}{2\,(m - 1)} \left(2\,\frac{r_i^2}{r^2}\right) = \sigma_x \cdot \frac{r_i^2}{r^2} \tag{9-17}$$

Its maximum value τ_{max} is attained (as long as $\sigma_t > \sigma_r$) at the contour when *r* is at a minimum, i. e., when $r_{min} = r = r_i$:

$$\tau_{max} = \sigma_x = \frac{2\sigma_x}{2} = \frac{\sigma_{t\,max}}{2} \tag{9-18}$$

9-4. Secondary Stress Conditions in the Plastic Zone

In that instance when the elastic deformations of a rock are exceeded, the foregoing assumptions do not apply any more; our knowledge about the real behavior of the rock is confined almost only on qualitative reasoning. In such a case, one usually makes use, then, of the theory of plasticity, especially of the Coulomb-Mohr failure condition. The reasoning may be such as set forth. If, in the rock, rupture or fracture occurs forming cracks and fissures, then in the ruptured rock around the shaft there sets in a plastic zone, or zone of stress release, called also the zone of *stress relaxation*. The relaxation of stresses decreases with increase in distance away from the shaft.

The stress release brings about a loosening of the rock, whereby the loosened rock material tends to intrude freely radially inward into the interior of the shaft. Because of this stress relaxation effect, the cross-section of the shaft is reduced. A lining of the wall of the shaft takes up the rock pressure, thus establishing a new condition of equilibrium, and the stress-relaxed zone extends out radially and terminates at a distance r_0 from the vertical axis of the shaft (Fig. 9-3).

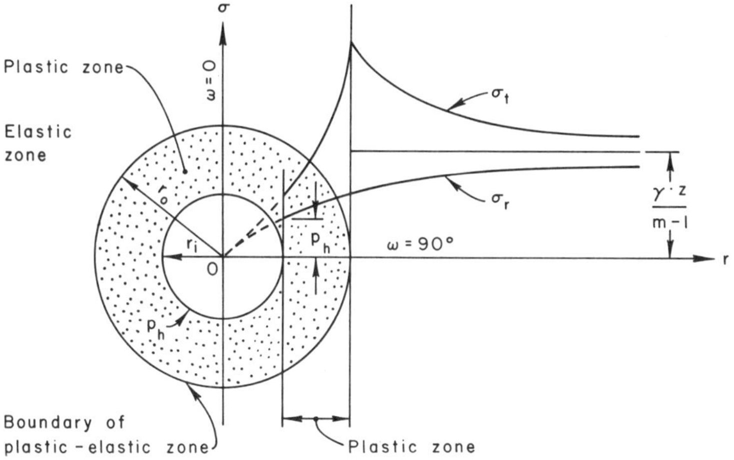

Fig. 9-3
Theoretical course of stress distribution in plastic and elastic zones of rock around a vertical shaft

In this figure, p_h is the lateral pressure exerted on the outside mantle surface of the vertical, cylindrical lining of the shaft or borehole, whichever the case is.

The pressure p_h on the lining from granular material is given as:

$$p_h = \frac{\gamma \cdot z}{m-1}(1 - \sin\varphi)\left(\frac{r_i}{r_o}\right)^{\frac{2\sin\varphi}{1-\sin\varphi}} \tag{9-19}$$

This pressure attains a maximum when the plastic zone reaches the lining of the shaft, i.e., when $r_o = r_i$. Then the pressure is:

$$p_{h_{max}} = \frac{\gamma \cdot z}{m-1}(1 - \sin\varphi) \tag{9-20}$$

The pressure decreases with increase in r_o. Equation (9-20) also shows that pressure decreases with increase in φ.

The stress components are:

$$\sigma_r = \left[\frac{\gamma \cdot z}{m-1}(1 - \sin\varphi)\right]\left(\frac{r}{r_o}\right)^{\frac{2\sin\varphi}{1-\sin\varphi}} \tag{9-21}$$

$$\sigma_z = \gamma \cdot z \tag{9-22}$$

$$\sigma_t = \left[\frac{\gamma z}{m-1}(1 + \sin\varphi)\right]\left(\frac{r}{r_o}\right)^{\frac{2\sin\varphi}{1-\sin\varphi}} \tag{9-23}$$

These equations are valid only when in the plastic zone the value of the third principal stress σ_z lies between the values of the other two principal stresses σ_r and σ_t.

To satisfy these equations, POISSON's number should be in the order of magnitude of 2. This means that the validity of such computations is severely limited, because in practice there are no rock materials which would deform without change in volume.

For solid rocks in the plastic zone, the following relationship can be written:

$$\sigma_t - \sigma_r = \sigma_{unc} \tag{9-24}$$

where σ_{unc} = unconfined strength of the rock.

The lateral pressure p_h in the plastic zone for solid rock calculates as:

$$p_h = \frac{\gamma \cdot z}{m-1} - \frac{\sigma_{unc}}{2}\left(2 \ln\frac{r_o}{r_i} + 1\right) \tag{9-25}$$

Within the region of $r_i < r < r_o$ at a distance r from the axis of the shaft, the stress components are:

$$\sigma_r = \frac{\gamma \cdot z}{m-1} + \frac{\sigma_{unc}}{2}\left[2\,ln\left(\frac{r}{r_o}\right) - 1\right] \tag{9-26}$$

$$\sigma_z = \gamma \cdot z \tag{9-27}$$

$$\sigma_t = \frac{\gamma \cdot z}{m-1} + \frac{\sigma_{unc}}{2}\left[2\,ln\left(\frac{r}{r_o}\right) + 1\right] \tag{9-28}$$

These three equations describe the stress conditions in the plastic zone.

At the lining or casing, by Eq. (9-25),

$$p_{h_{max}} = \frac{\gamma \cdot z}{m-1} - \frac{\sigma_{unc}}{2} \tag{9-29}$$

In contradistinction to granular material, the plastic zone in solid rock is limited. The larger the cohesion, the smaller is the zone where the material is plastic.

All these calculations show that the prevailing stresses in the individual strata are considerably altered when sinking a shaft, and that these changes in stress conditions are independent of the radius of the shaft or borehole.

Borings are usually penetrated into rock to a much greater depth than tunnels are driven.

This favorable circumstance (the depth) is not merely because the boring diameter is much less than that of a tunnel, but also because the horizontal tunnel, or gallery, is subjected to the vertical surcharge load of the superimposed rock, whereas in the case of the borehole the lesser horizontal stresses bring about a rotational-symmetrical stressing of the rock around the borehole.

9-5. Other Forms of Openings and Modes of Stress Distribution in Rock

For analyses of other forms than circular openings — such as elliptic, parabolic, horseshoe-like, square, and rectangular ones — refer to FENNER (3), TERZAGHI and RICHART, JR. (27), TERZAGHI (26), KASTNER (16), SZÉCHY (24), for example.

For surface loadings of an elastic medium by concentrated single and line loads; uniformly distributed; triangular and circular flexible surface loading giving influence values for stresses in the elastic medium at any point; and for triaxial, central-symmetrical vertical stress from rigid circular surface loading, refer to the author's publications, References 7-15.

CHAPTER 10

SOME ENGINEERING PROBLEMS ASSOCIATED WITH WORK IN ROCK

10-1. Kinds of Problems in Rock Engineering

Design in rock commonly comprehends structures as:

1. Foundations and anchorages;
2. Rock slopes; and
3. Various subsurface underground excavations: tunnels, shafts, mines, adits, and the like.

In underground rock engineering, the activities involved are:

1. Breaking and removing rock, and
2. Maintaining control over surrounding rock.

Therefore, the basic engineering science for underground as well as surface rock engineering is the discipline of rock mechanics.

Rock pressure, mechanical and chemical action of the underground water, pore-water pressure, temperature variations, frost action, gases, prestressing forces, and tectonic forces—they all impart their effects on underground structures, as well as on tunnel linings (1, 4, 18-24, 26, 28, 30).

In rock engineering, one encounters several possible problems called by Terzaghi (26) hazards.

Economically, hazards in underground opening works constitute unanticipated contingencies such as expenses and delay in such works, loss of materials, and sometimes also loss of life.

Usually hazards are caused by the departure in performance of the rock from the statistical average of similar materials, affecting seriously the state of its structure.

Generally, the causes of hazards are:

1. Abnormally high rock loads and thus high natural, primary stresses in rock;

2. Unfavorable rock conditions, such as rock defects, and orientation of stratification, joints, fissures, and underground cavities;

3. Excessive influx of water into an underground opening;
4. Temperature conditions in the interior of the rock massif;
5. Presence of harmful gases;
6. Seismic activity; and
7. Other possible unforeseen causes, among them chemical reaction of pore water on rock minerals, and radioactivity, for example.

Obviously, such hazards depend on the regional geology and the *en masse* properties of rock materials under consideration. Thus, in a way, in geotechnical engineering such hazards mean a financial risk.

Rock Pressure

Rock and soil pressures, and water pressure on underground structures at great depths, are the most important potential loads. The determination of the magnitude of rock pressure is one of the most intricate problem in rock mechanics.

There are forces prevailing in the rock, the causes of which reside in the deeper formations of the earth. These forces have brought about stresses in the earth's crust, bringing about plastic deformations as well as fractures in the rock. But even now, the earth's crust has not come to rest, but is always in motion. Tectonic and orogenic rock-forming processes are in progress even today.

It is, therefore, impossible to comprehend the rock-forming forces. In order to determine them, measuring instruments would have to be installed underground. However, to do this, it would be necessary to make a hole, an opening, or a borehole, or a shaft in the ground. To do this, the stress conditions in the rock would be readily disturbed. The results of in-situ measurements are also very much affected by tectonics of the location and by the method of measurement.

Therefore, in analyzing stress conditions in rock around a vertical, circular shaft, one usually assumes that the effects of rock-forming forces on rock have already faded out, and only the effect of the gravity force is considered.

The rock structure, depending upon its age and position of formation, is not homogeneous; that is, the composition and therefore, also its mechanical properties are not the same in all directions. The transformation of the original deposition which the rock has gone through—such as folding, faulting, and the natural inferiorities of the rock—all these, too, influence its mechanical behavior.

In rock engineering, many cases are known where the occurrence of plastic phenomena in, and fracturing of rock around a relatively narrow zone of underground openings sometimes cannot be satisfactorily explained by the prevailing overburden surcharge load—whether at great depth, at the ground surface or, especially, when near the ground surface.

10-2. Inferiorities of Rocks

In rock massifs, seldom, if ever, can ideal conditions be expected. Because of their presence, rock joints are important factors in excavating rock. Open, continuous joints form passages for groundwater circulation in rock.

Closely spaced joints impair stability of underground structures and foundation excavations in rock and rock slopes. In tunneling work, joints contribute to overbreak.

Working below the groundwater table, open joints are usually the source of water seepage into the underground opening, thus contributing to difficulties in work, as well as to added expenses for coping with the seepage by pumping and/or drainage facilities. Also, rock weathering, because of water, is a very undesirable factor.

Knowledge of fault and joint locations and their orientation is also of particular importance in rock engineering. Their extent also can lead to water problems, because joints can act as channels for conveying water through relatively dense strata.

Gouge is of the utmost concern in rock engineering. In underground openings, gouge in faults and joints may create disastrous hydrostatic head, viz., pressure.

Defects which intersect any kind of rock in situ may cause a very large scatter of strength values depending upon the number, character, distribution, and location of the various kinds of rock defects, and on the volume of the domain involved.

Natural and man-made activities in rock can develop into large-scale surface settlements or local sinkholes.

10-3. Subsurface Water

There is more than one effect of subsurface water on an underground structure. Groundwater finds its way toward a tunnel, bringing about erosion, washouts, and/or silting of tunnel drainage facilities (23).

Water dissolves certain chemicals, and some of them may be deleterious to concrete. Also, frost and its bursting action may contribute damage to a structure. Serious water problems may arise in the construction and maintenance of tunnels intersecting synclines containing water-bearing strata. Therefore, determination of physical characteristics of the water-bearing materials and their structure is of paramount importance.

Driving of tunnels in built-up urban areas may bring about dangerous lowering of the groundwater table, causing subsidence of streets and basement floors as well as rotting of timber piles, thus imparting distress to buildings. Drying-out of wells near caverns and tunnels is a well-known phenomenon. Underground openings are drainage centers. Relatively small water seepage into underground openings cause considerable subsidence, viz., settlement.

10-4. Temperature

From measurements made in underground openings and deep borings it is known that there exists a geothermal gradient in the earth's crust from the interior to the exterior of the earth. This means that there exists a terrestrial heat flux from the hot interior of the earth toward the cooler ground surface. This also means that the thermal regimen in the upper courses of the earth depends to a certain degree upon the geothermal gradient of the rock and/or soil materials.

The geothermal gradient varies greatly from place to place, depending upon the geologic history of the region, the kind of material, and stratification of the rock. Underneath mountain ranges, the average geothermal depth step is 35 to 60 m/°C, and underneath valleys it is 20 to 25 m/°C.

Some extreme geothermal depth steps are:

Comstock mine (U.S.A.)	17 m/°C

Canadian Shield	125 m/°C
Gold ore mines, Johannesburg	110-130 m/°C

The thickness of the overburden cover of geological materials above a tunnel likewise influences the temperature in tunnel tubes already in service. The temperatures in some tunnels are compiled in Table 4-4.

High temperatures in a tunnel cause uncomfortable, ineffective work, and need air-conditioning and ventilation.

Heat, positive and negative (freezing), also induces thermal stresses in the rock. Thermal stresses occur where a body (say, rock) at its various points is warmed or cooled at various heat intensities. The cause of thermal stresses is the cooling of the magmatic rocks, or warming of rock which formerly was covered with a sheet of ice.

If underground caverns in rock are designed for storage of liquefied gas, great problems arise at very low temperatures (< -100 °C). Upon freezing of the groundwater, the low temperature causes the opening up of new cracks in the rock. The ice formed in the joints and cracks has no sealing effect because the rock contracts as the cooling proceeds, causing opening up of the ice-filled joints. According to MORFELDT (18), the contractions are considerable, more than 0.1 m per 100 m of cavern.

An increase in temperature lowers rock strength and increases ductility. In compression tests of granite, GRIGGS et al. (4) found nearly a threefold decrease in strength as temperature was raised from 25 ° to 800 °C. Basalt showed nearly an eightfold decrease over the same temperature range. Dolomite showed a twofold decrease and marble an eightfold decrease in strength.

10-5. Hazards in Various Kinds of Rocks

As to geology, underground hazards can be encountered in igneous rocks, sedimentary rocks, and metamorphic rocks (1, 26, 30).

Thus, a thorough rock exploration is imperative before any rock engineering design and construction are contemplated.

Igneous Intrusive Rocks

Generally, the probability of underground hazards in igneous instrusive rocks, such as sound granite, is relatively small and of rare occurrence. However, deviations from the normal are always possible (30). Even in the best granites in Maine, plastic flow has been found in the rock at some 150-ft depths. "In one case, the solid sheets buckled upward with serious pop-outs and the fractures made further excavation wasteful since so little usable stone was recovered," according to Feld (2).

The possibility of treacherous conditions that might exist in chemically altered igneous rocks is difficult to ascertain at depths from the surface.

Also, flaking, popping, and spalling of rock exposures in deep tunnel construction should be anticipated (28). In the French part of the Mont Blanc tunnel, over 100,000 rock anchors were inserted to prevent rock bursts in the overstressed granite.

Igneous Extrusive Rocks

Breccia and tuff formations may be found to be in a state of decomposition. Because of fault zones, water finds its way into the underground openings made in such rocks. The presence of soft breccia may cause movement in a tunnel that intersects a fault. In porous basalt, large quantities of water may be present.

Because of the porosity of extrusive rocks, harmful gas movement is quite common. Hence, such rocks present themselves as hazardous.

Sedimentary Rocks

Limestone

Usually, limestone and sandstone above a water table present relatively few problems. However, deformations and cracks caused by actual overburden pressure are most likely to occur in limestones.

Limestones as a foundation for a dam fail because of hydration along leaking cracks.

Frequently, in limestone, solution cavities are encountered. These, and also fissures, may be fully or partly filled with fine-particled soil or they may be empty.

Considerable leakage through the limestone can be expected, especially if a cavernous limestone is situated under a dam or in a water-impounded reservoir. Also, limestone may contain thin layers of sandstone. These sandstone layers are permeable, and are weaker than limestone. A large-pore or cavern-containing limestone has a very low compressive strength ($< 211\ kg_f/cm^2 \approx 3000$ psi).

Any man-made change introduced in underground conditions in the limestone-dolomite rocks should be taken as a warning of possible subsidence of such rocks, as well as change-of-water circulation that will dissolve the rock.

Sometimes limestone grains are cemented together with clayey material. The cementing strength may become reduced upon wetting.

Shale

The predominant characteristic of a shale is the high percentage of fine clay particles, and also silt. The strength of shales varies from that of a sound shale rock to that of a swelling clay. Where porous, water-bearing formations rest on shale, a large influx of water from the porous formation into the shale can take place.

Shales disintegrate easily, and so do sandstones and many slates.

Buried seams in rock filled with saturated gouge may become liquefied even by a minor earth tremor and may bring about a major rock slide. Shales may also contain thin seams of plastic clay. Slides on buried seams in the rock may be the cause of catastrophic dam failures. Elastic rebound and plastic deformations in shale are frequently observed phenomena, also.

With large overburden pressure, shale is to be regarded as hazardous.

Schist

The hazards in sound schist are generally rather mild. The tall PAN-AM Building over the Grand Central Railroad Station in New York City is founded on what is called Manhattan schist (a mica schist). In altered schists, however, a large influx of water into underground excavations may take place.

Gneiss

Fine-grained gneiss may have open internal seams, frequently filled with water-deposited soft mud, seriously affecting the gneiss' engineering properties.

Requirements in the Design in Rock

Just as other engineering design must meet certain design requirements, design of structures on and in rock is also subject to the basic requirements, namely:

1. Structures in rock should be of safe design.
2. They should be purposely and safely constructed.
3. They should be esthetical in appearance.
4. They should be economical to build.

REFERENCES

1. Bolt, B. A. et al., 1975, *Geological Hazards*. New York, N. Y.: Springer-Verlag.

2. Feld, J., 1966, *Rock as an Engineering Material*. Evanston, Ill.: Soiltest, Inc., p. 18.

3. Fenner, R., 1938, "Untersuchungen zur Erkenntnis des Gebirgsdrucks," *Glückauf*, No. 32.

4. Griggs, D.T. et al., 1960, "Deformation of Rocks at 500° to 800°C," *American Geological Society Memoir 79*, pp. 39–104.

5. Heim, A., 1912, "Zur Frage der Gebirgs- und Gesteinsfestigkeit," *Schweizerische Bauzeitung*, Vol. 24, February 1912.

6. Horvath, J., 1965, "A New Approach to the Determination of Stresses in the Earth's Crust and Strata Pressure on Tunnel Linings," *International Journal for Rock Mechanics and Mining Science*, Vol. 2, pp. 327–340.

7. Jumikis, A. R., 1956, "Rupture Surface in Sand under Oblique Loads," *Proceedings ASCE, Soil Mechanics and Foundations Division*, Vol. 82, Paper 861, pp. SM–1 to SM–6.

8. Jumikis, A. R., 1961, "The Shape of Rupture Surface in Dry Sand," *Proceedings Fifth International Conference on Soil Mechanics and Foundation Engineering*, held July 7–22, 1961. Paris: Dunod, Vol. 1, Paper 3A/23, pp. 693–698.

9. Jumikis, A. R., 1962, *Soil Mechanics*. Princeton, N. J.: D. Van Nostrand Company, Inc., pp. 641–651.

10. Jumikis, A. R., 1965, *Stability Analyses of Soil-Foundation Systems: A Design Manual (Based on Logarithmically Spiralled Rupture Curves)*. New Brunswick, N. J.: Rutgers University, College of Engineering, Bureau of Engineering Research, Engineering Research Publication 44 (contains 13 tables for the construction of logarithmically spiralled rupture surface curves).

11. Jumikis, A. R., 1969, *Theoretical Soil Mechanics*. New York, N. Y.: Van Nostrand Reinhold Company, pp. 41–55.

12. Jumikis, A. R., 1973, *Settlement Tables for Centrically Loaded, Rigid Circular Footings on Multilayered Soil Systems*. New Brunswick, N. J.: Rutgers University, College of Engineering, Bureau of Engineering Research, Engineering Research Bulletin 54.

13. JUMIKIS, A. R., 1973, "Settlement Influence-Value Chart for Rigid Circular Foundations." Washington, D. C.: *Highway Research Board Record,* No. 457, National Research Council - National Academy of Sciences - National Academy of Engineering, pp. 27–38.

14. JUMIKIS, A. R., 1973, "Vertical Stress Chart for Rigid Circular Foundations," *Proceedings ASCE, Soil Mechanics and Foundations Division,* Vol. 99, No. SM 12, Paper 10189, pp. 1196–1201.

15. JUMIKIS, A. R., and A. A. JUMIKIS, 1975, *Red Brunswick Shale and its Engineering Aspects.* New Brunswick, N. J.: Rutgers University, College of Engineering, Engineering Research Bulletin 55.

16. KASTNER, H., 1971, *Statik des Tunnel- und Stollenbaues auf der Grundlage geomechanischer Erkenntnisse.* Berlin: Springer-Verlag.

17. MINDLIN, R. D., 1939, "Stress Distribution around a Tunnel," *Proceedings ASCE, Soil Mechanics and Foundations Division,* April 1939, pp. 619–642.

18. MORFELDT, C. O., 1970, "Significance of Groundwater at Rock Construction of Different Types," in *Large Permanent Underground Openings.* Oslo: Universitetsforlaget, pp. 305–317.

19. OBERT, L., and W. I. DUVALL, 1967, *Rock Mechanics and the Design of Structures in Rock.* New York, N.Y.: John Wiley and Sons, Inc., p. 495.

20. PALMER, J. H. L., and K. Y. LO, 1976, "In Situ Stress Measurements in Some Near-Surface Rock Formations-Thorold, Ontario," *Canadian Geotechnical Journal,* Vol. 13, No. 1, pp. 1–7.

21. ROCHA, M., 1965, *Some Problems on Failure of Rock Masses, Memória No. 258.* Lisbon: Laboratório Nacional de Engenharia Civil.

22. ROŠ, M., and A. EICHINGER, 1949, *Die Bruchgefahr fester Körper.* Zürich: Eidgenössische Materialprüfungs Anstalt- EMPA Report No. 172.

23. SERAFIM, J. L., 1968, "Influence of Interstitial Water on the Behaviour of Rock Masses," in *Rock Mechanics in Engineering Practice.* London: John Wiley and Sons, Inc., Chapter 3, pp. 55–97.

24. SZÉCHY, K., 1966, *The Art of Tunnelling.* Budapest: Académiai Kiadó, pp. 83–84.

25. TALOBRE, J. A., 1957, *La Mécanique des Roches.* Paris: Dunod.

26. TERZAGHI, K., 1946, "Tunnel Hazards," in *Rock Tunneling with Steel Supports,* by R. V. PROCTOR and T. L. WHITE. Youngstown, Ohio: The Commercial Shearing and Stamping Company, pp. 96–99.

27. TERZAGHI, K., and F. E. RICHART, Jr., 1952, "Stresses in Rock about Cavities,:" *Géotechnique* (London), Vol. 3, No. 2 p. 57.

28. TERZAGHI, K., 1962, "Dam Foundation on Sheeted Granite," *Géotechnique* (London), Vol. 12, No. 3, p. 199.

29. TIMOSHENKO, S., and J. N. GOODIER, 1951, *Theory of Elasticity.* New York, N. Y.: McGraw-Hill Book Company, Inc., pp. 78–81.

30. WHITE, W. S., 1946, *Rock Bursts in the Granite Quarries at Barre, Vermont,* U.S. Geological Survey Circular No. 13 (15 pages).

PART 4

STABILIZATION OF ROCK

CHAPTER 11

ROCK REINFORCEMENT

11-1. Anchoring

In general, rock bolt support or rock anchorage is replacing or supplementing conventional timber and steel supports for the roofs and sides of rock in underground openings (5, 8). Bolt support or rock anchorage is also used for stabilizing rock slopes and mines. Rock anchoring is used, for example, for reinforcement of arch-dam rock abutments, caverns, and other rock support in foundation, tunnel, and hydraulic structures engineering (Figs. 11-1 through 11-4). Whenever a naturally solid, self-supporting roof is encountered above a false hanging wall or weak layer, the wall can be held up by a bearing plate and bolts anchored in the solid rock formation above (Fig. 11-1).

11-2. Rock Bolting

To avoid failure of structures made in rock and rock slopes, rock reinforcement is used. Rock reinforcement is a means by which strength is added to the rock to avoid its failure. The reinforcing elements are steel rods, rock bolts, or stressed steel cables. Rock bolting, or rock anchor, is used for reinforcing rock slopes to prevent rock blocks from falling away from the main rock mass when isolated by joints and faults, or sliding down along an inclined plane of rock. Hence, the reinforcement has for its purpose to load the rock in such a manner that the resistance to shear along planes of weakness is increased.

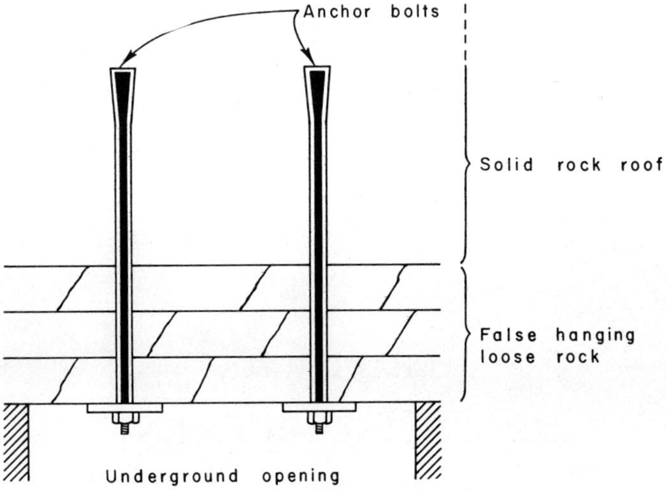

Fig. 11-1
Anchoring a weak zone of rock to a stronger zone with rock bolts and bearing plates

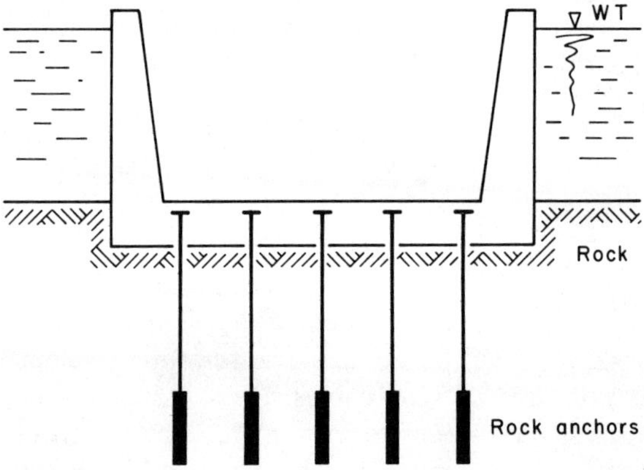

Fig. 11-2
Anchoring of a U-shaped structure against hydrostatic uplift

a) Failure in shear of sand under an inclined load, $\varphi = 30°$ (Author's study, Ref. 7)

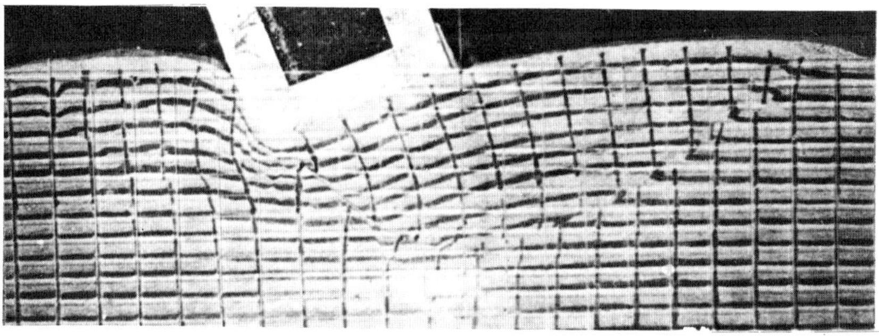

b) Anchoring of a potential rupture wedge in rock against shear and subsequent lateral expulsion from undernearth the base of a foundation

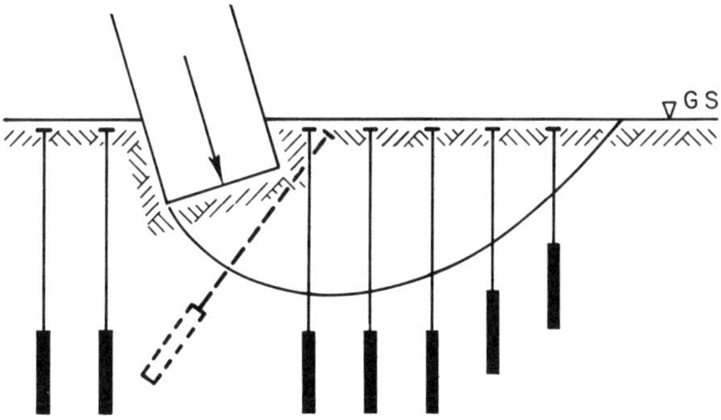

Fig. 11-3
Curved rupture surfaces in soil and rock

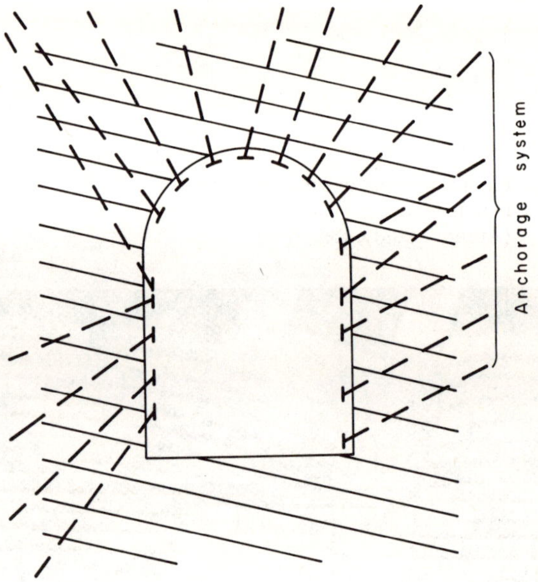

Fig. 11-4
Rock bolting in fissured rock around an underground opening

Requirements of Anchorage

1. The anchorage must be strong.

2. The rock must be *strong* and *continuous* on a limited scale to provide a firm anchorage for the bolts. However, anchorages may also be provided by cement grouting.

3. The bolts should be of adequate length to create a precompression zone of sufficient extent around, or along, a structure in rock to cope with stresses of failure. Rock bolts should be anchored beyond the tension zone.

11-3. Effect of Rock Bolting on Shear Stress

In reinforcement of jointed rock, the effect of shear strength of the joint on the design of rock bolts and anchors enters into consideration. Upon introduction of the reinforcement, precompression is applied to the solid rock through tensioning, viz., prestressing of the anchor bolt. Before prestressing, the ends of the rock bolts, or anchors, should be firmly concreted into the borehole of the anchor.

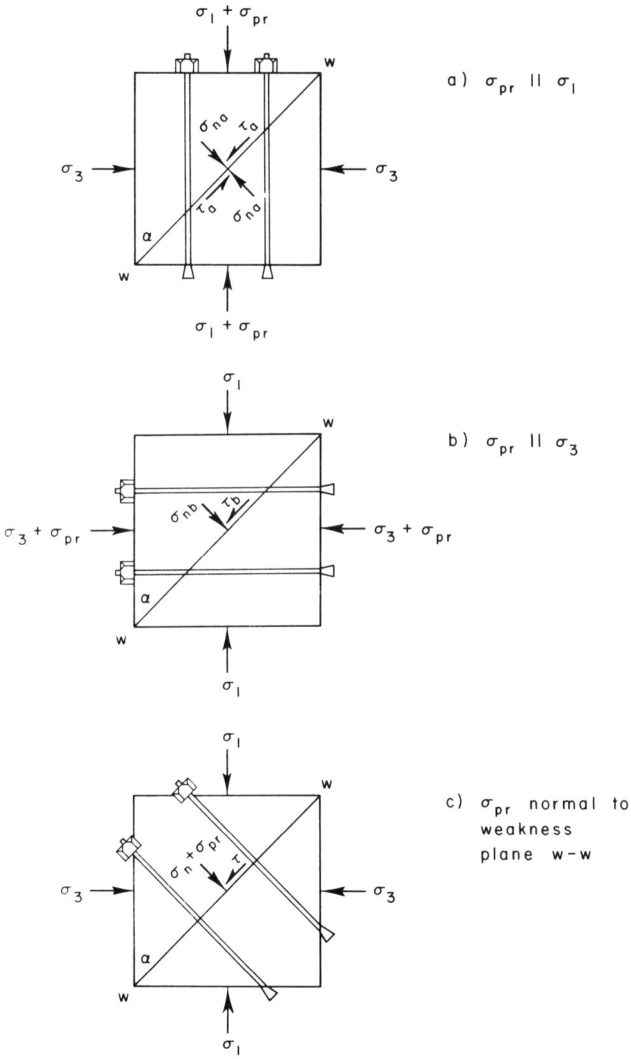

Fig. 11-5
Imparting precompression stress σ_{pr} on bolted rock elements
(After FARMER, Ref. 32)

As shown by FARMER (3), a precompression stress, σ_{pr}, can be imparted to the element of a rock, separated by a diagonal plane of weakness, in one of the three different ways as sketched in Figs. 11-5a, b, and c:

1. Precompression σ_{pr} applied parallel to the major principal stress σ_1;
2. Precompression σ_{pr} applied parallel to the minor principal stress σ_3, and
3. Precompression stress σ_{pr} applied normal to the shear plane, viz., plane of weakness.

The maximum load P per bolt is computed as

$$P = \sigma_{ult} \cdot a \tag{11-1}$$

where σ_{ult} = ultimate compressive stress of the rock, and
a = cross-sectional area of the anchorage.

The applied precompressive stress σ_{pr} to rock is:

$$\sigma_{pr} = \frac{n \cdot P}{A} \tag{11-2}$$

where

n = number of bolts, and

$A = \sum_1^n (a)$ = total area of the bolted rock.

Using the normal- and shear-stress equations

$$\sigma_n = \frac{\sigma_1 + \sigma_3}{2} + \frac{\sigma_1 - \sigma_3}{2} \cos 2\alpha \tag{11-3}$$

$$\tau = \frac{\sigma_1 - \sigma_3}{2} \sin 2\alpha \tag{11-4}$$

and adjusting them for the prestressing condition, obtain normal and shear stresses on the shear plane:

Case (a), $\sigma_{pr} \parallel \sigma_1$:

$$\sigma_{na} = \sigma_n + \frac{\sigma_{pr}}{2} \cdot (1 + \cos 2\alpha) \tag{11-5}$$

$$\tau_a = \tau + \frac{\sigma_{pr}}{2} \cdot \sin 2\alpha \tag{11-6}$$

Case (b), $\sigma_{pr} \parallel \sigma_3$:

$$\sigma_{nb} = \sigma_n + \frac{\sigma_{pr}}{2} \cdot (1 - \cos 2\alpha) \tag{11-7}$$

$$\tau_b = \tau - \frac{\sigma_{pr}}{2} \cdot \sin 2\alpha \tag{11-8}$$

Case (c), σ_{pr} normal to weakness plane $w - w$:

σ_{nc} and τ_c to be evaluated by Eqs. (11-3) and (11-4).

Furthermore, FARMER (3) shows that the COULOMB shear strength line or envelope for $\alpha = 45°$ for the above mentioned three cases are:

$$\text{a) } \tau = \sigma_n \cdot \tan\varphi - \frac{\sigma_{pr}}{2}(1 - \tan\varphi) + c \tag{11-9}$$

$$\text{b) } \tau = \sigma_n \cdot \tan\varphi + \frac{\sigma_{pr}}{2}(1 + \tan\varphi) + c \tag{11-10}$$

$$\text{c) } \tau = \sigma_n \cdot \tan\varphi + \sigma_{pr} \cdot \tan\varphi + c \tag{11-11}$$

Observation of these equations shows that there will be a weakening in the shear strength τ of the rock structure in case a) and strengthening in cases b) and c). The maximum strengthening is obtained in the case (b) when precompression aids the minor principal stress σ_3, which is the lateral confining stress.

11-4. Rock Bolt Support

After excavation of an underground opening in rock, the stresses of a rock mass around and near the opening and/or excavation tend to become relaxed. To maintain the initial, natural, tightly packed state of the rock mass, rock bolts are used.

Rock bolts not only take care of excessive relaxation of the rock mass around the contour of the opening, but they can also prevent a sudden decompression of the rock material brought about by rock spalling and rock bursts. To avoid this, the bolts must be installed immediately upon and after making the excavation of the opening, viz., driving of the tunnel. Rock bolts prevent massive falls of loose rock fragments, thus supporting roofs and ceilings of underground openings in rock, and slope faces of rock in open excavation.

The performance of rock-bolted layers of rock formations resembles that of a bolted beam. The rock layers are usually tied together by means of pretensioned rock bolts. On the average, the maximum spacing of the bolts is taken as approximately one-half on the length of the bolt. Figure 11-6 is a schematic representation of some rock-bolt support systems.

In principle, a rock bolt is installed as follows.

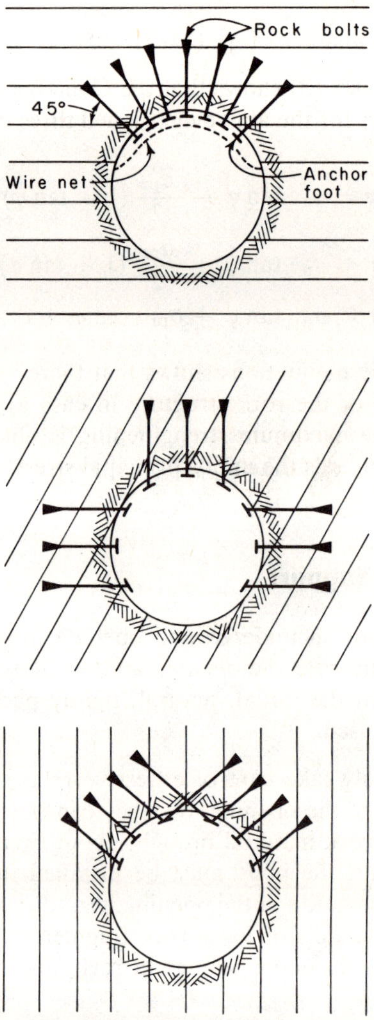

Fig. 11-6
Some rock-bolt support systems

A hole is drilled through the unstable rock into a stable one, into which hole a long steel bolt is inserted. With the appropriate head-wedge assembly employed, the bolt is securely fixed and keyed in the hole. In this way, the outer layer of a rock slab is bolted to the stable layer in which the bolt is keyed.

To protect against falling of smaller rock fragments from the ceiling, the feet of the anchors or bolts around and along the vault-like ceiling are covered with a wire net.

The most suitable form of rock bolting, bolt tensions, and rock displacements can be determined by appropriate tests.

As to the types of rock bolts and anchors, there are several types of them on the market. Almost every rock engineering firm which is in the business of rock anchorage has its own trademarked and/or patented anchor design. The type of rock bolts and rock anchors to use depends upon the nature of the construction site, the geology of the rock and its properties, the load to cope with, the method of installation of the anchors, and possibly other factors. Each anchor type has its place and use in foundation, hydraulic structures, an rock engineering. To choose the type of anchor to use, one would refer to anchor manufacturers' catalogs.

11-5. Determining Rock Bolt Support

The analytical delineation, viz., thickness of the tensile stress zone in rock above the crest of a circular opening, is discussed in Section 7-8. This delineation enables one to calculate the weight of the loosened rock, also called the gravity or overburden load of the rock mass in the upper part of an underground opening. The thickness of this loose, fragmented rock mass depends upon the quality of the rock, of course, as well as upon the construction method in the rock and the kind and method of rock support used against caving-in.

One method to stabilize the loose rock above the crest of the circular opening, and thus to safeguard against rock fragments falling into the opening, is the one known by the term "rock-bolt support."

As shown by CORDING (1, 2), the thickness t of the zone of the loosened rock above the crest of the circular opening in rock comprising the gravity load W (weight of the loosened rock) may be determined empirically as

$$t = n \cdot D \tag{11-12}$$

where t = thickness of the rock load on the crest of the opening (Fig. 11-7),
n = a function of support rigidity and strength of the rock mass (10), and
D = diameter of the circular underground opening in rock.

Equation (11-12) thus indicates that the thickness $(n \cdot D)$ of the rock load depends upon (1) the rock quality, (2) the kind of support and construction method used, and (3) the width, viz., diameter, of the opening. As indicated by CORDING, the value of n will be small for an excellent-quality rock if the rock is rapidly reinforced and supported with rock bolts. In such a case, the bolt pressures p_i on rock necessary to support the gravity load W will be relatively small. For example, if bolt pressures in the dome of a cavity were 20 psi = 137.9 kN/m², this pressure would support a 20-foot-thick (6.096 m) gravity-loaded rock (n = 20 feet/100 feet = 0.2 = 6.096 m/30.48 m for this case). The bolt pressures p_i needed for stability increase in proportion to the width (diameter) of the opening.

If the cohesion of the rock mass is small (visualize a heavily jointed rock mass around a large opening), then the pressure p_i required for stability, as given empirically by CORDING (1), is

$$p_i = n \cdot D \cdot \gamma \qquad (11\text{-}13)$$

where γ = unit weight of the rock mass.

If the size of the underground opening is small with respect to the fractures and cohesion must be considered, the bolt pressure p_i on rock is

$$p_i = n \cdot D \cdot \gamma - n' \cdot c \qquad (11\text{-}14)$$

where n' = a function of the rock mass quality and the kind of support method, and
c = cohesion per unit area of the rock mass.

In a small tunnel driven through a sound, excellent-quality rock, the cohesion of rock is usually large enough so that no support (= no p_i) is needed.

The opening must also be safeguarded against movement of the most critical rock wedge into the opening (Fig. 11-8). The normal pressure P_N tends to drive the wedge into the opening (1). The necessary bolt pressure p_i on the rock for preventing failure of the wedge depends upon the total shear resistance S along the failure plane. Low shear resistance may be incurred because of a high hydrostatic pressure U on the failure plane, or because there is present, on the potential plane of failure, a gouge material with a low frictional strength.

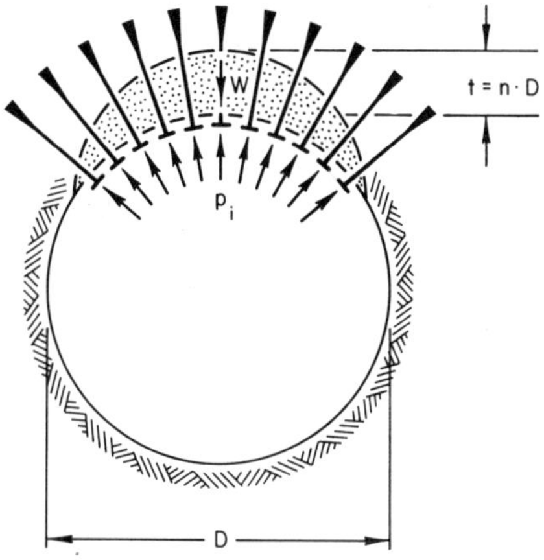

Fig. 11-7
Support of gravity load

W = weight of loosened rock
p_i = internal pressure applied by bolts
n = function of support rigidity and strength of rock mass

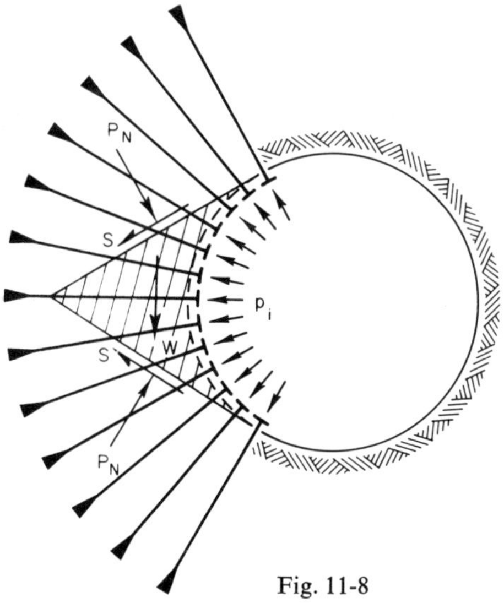

Fig. 11-8
Support of critical wedge
(After CORDING, Ref. 1)

$S = (P_N - U) \cdot \tan\varphi + C$
U = groundwater pressure
φ = angle of friction
C = cohesion
W = weight of wedge

The shear resistance, or shear strength S, is written as

$$S = (P_N - U) \cdot \tan \varphi + C \tag{11-15}$$

where U = hydrostatic or neutral pressure,
φ = angle of friction, and
C = total amount of cohesion in the shear plane.

The internal pressure p_i to support the worst rock wedge is thus a function of the normal force P_N and the size of the opening.

The lengths of the rock bolts should be long enough to extend them through the worst rock wedge. Usually bolt lengths are one-fourth to one-third of the span of the underground opening.

11-6. Some Advantages and Disadvantages of Rock Bolting

Although rock-bolting is not a cure-all and will not work in all cases where applied, it has several advantages over timber support methods for stabilizing rock masses. Some of the rock-bolting advantages are:

1. A better underground-opening roof, and an open-excavation slope support where other methods fail;
2. Economy;
3. Relatively short installation time of anchors and/or bolts;
4. Uncluttered storage space for materials;
5. Unobstructed working space (there are no "forests" of supporting timber; this facilitates a convenient use of mechanical equipment);
6. Relatively long duration in service, outlasting timber;
7. Improved ventilation in underground openings; and
8. Choice in direction and magnitude of the retaining forces, thus taking care of the various attitudes of the rock and the different orientations and magnitudes of shear strength of the various-quality joints.

Some of the disadvantages of rock-bolting are:

1. Poorer performance in extremely blocky and loose rock formations where no solid anchoring medium can be found, and
2. Impossibility of ascertaining the exact condition of the bolt in the bore-hole.

CHAPTER 12

ROCK SLOPES

12-1. Definitions

In highway, foundation, and hydraulic structures engineering in rock, there is a need for assessing the degree of stability of man-made as well as natural rock slopes.

Highway construction contributes to rock slides during excavation and after completion.

A slope is a vertical or inclined boundary surface between air and rock, or the body of an earthwork such as a dam, cut, or fill.

In geotechnics, the topic "stability of slopes" is dealt with from two engineering viewpoints, namely:

1. The design of man-made slopes of cuts and fills in advance of new rock or earthwork construction, with prescribed safety requirements against failure, and
2. The study of stability of existing slopes of rocks and earthworks, slopes which are potentially unstable, slopes which have failed, or slopes which have to be redesigned.

The concept of the "stability" of a slope is an indeterminate one, because no slopes made in or of rock or soil can be regarded as fully guaranteed for their stability during their service over a period of many years. Climatic, hydrologic, and tectonic conditions, and man's activities in the immediate and/or adjacent area of the structure, underground opening, or earthwork may bring about, years later, changes affecting the stability of man-made and natural slopes. In particular, one should not overlook the possibility of the rock and soil becoming saturated by water with time.

The term "failure of slope" refers to any slope instability that affects its performance, and man's operation in a natural, geologic environment.

In conjunction with man's work in rock, instability of rock slopes usually results from an unfavorable, critical attitude and the location and frequency of the various discontinuities of the rock formations in question. Conventionally, the concept of failure refers to the stress condition in rock, when the externally applied stress exceeds the strength of the rock. If this happens, the rock material breaks down, and the material is finished.

12-2. Factors Contributing to Slope Failure

Theoretically, by means of the geostatic pressure formula $\sigma = \gamma \cdot H$, an ideal, sound, unfissured rock, such as a granite, for example, whose uniaxial compressive strength is $\sigma_u \approx 2200$ $kg_f/cm^2 = 2.2 \times 10^4$ $t_f/m^2 = 215.75$ MN/m^2 and whose unit weight $\gamma = 2.6$ $t_f/m^3 = 25.5$ kN/m^3, would stand vertically at a height of $H = \sigma_u/\gamma = (2.2 \times 10^4)/2.6$ $t_f/m^3 = 215.750/25.5 \approx 8460$ m. However, in nature there is no ideal rock whose vertical slope can stand this high.

Observations in the field have shown that many gentle slopes of much less height have failed. This means that the critical height of a rock slope is governed not merely by the shear strength of the rock alone, but also by the various rock defects, such as jointing, cracks, fissures, and other possible weaknesses. Also, the physical characteristics of a rock mass, the mechanism involved in rock failure, the effect of water on rock — all these factors affect the engineering structures made in rock.

Continuous fissures and jointing systems separate the rock mass partly or completely into blocks. The shear strength along these planes of weakness is considerably low. Therefore they offer low resistance to displacement of the rock blocks, and consequently low resistance to failure. This is even more true if the orientation of the rock stratification system is inclined outward of the rock mass. Besides, through the jointing system water usually circulates. These conditions merely indicate that without special engineering measures, seldom are dry, leakproof, underground openings and slopes in rock encountered. Gravity, weathering, and erosion brought about by a humid climate are factors contributing eventually to the instability of rock slopes. Exposed to weathering effects directly, vertical and steep slopes of rock weather relatively quickly. Also, water flowing out from rock slopes influence, in time, their stability adversely. To this, frost action adds its share to the instability of rock slopes.

Slope design in rock and the stability of slopes of walls of rock excavations depend upon the following main factors:

1. The type of rock of which or in which the slope is made;
2. The structure, stratification, and attitude of the rock formations (angle of dip of rock strata, for example);
3. Presence of a potential failure surface in rock (frequency of geological discontinuities) and the steepness of its angle of dip toward the excavation or valley;
4. The geometry of the cross-section of the slope (height, slope angle, berms, for example);
5. The degree of shattering of unsound rock in the slope;
6. The presence of zones of breccia, if any;
7. Unit weight of the rock slope material (gravity is one of the principal causes of all rock slides);
8. Magnitude of externally applied loads on the rock slope structure;
9. Distribution of weight and loads;
10. In-situ strength properties of slope rock (yielding of rock; creep; breaking, crushing, and rock-bursting of the rock mass brought about by high-intensity loads and stress concentrations in rock and by excavation-induced rock deformations);
11. Loosening up and separation of rock strata and sliding;
12. Loosening of rock and subsequent heaving because of stress relaxation in rock brought about by removal of rock constraint;
13. Loosening of rock strata from blasting;
14. The initial or primary stress regimen (stress field) in a rock mass before excavation;
15. Distribution of secondary, induced stresses in rock upon and after excavation;
16. Position of groundwater table;
17. Amount of moisture content (degree of saturation) in the rock (water is one of the most aggressive factors contributing adversely to the instability of slopes);
18. Vibrations and seismic forces;
19. Various environmental conditions and processes sculpturing the face of slopes (rock-weathering, frost, and chemical action of pore-water on rock minerals).

All of the above-listed factors (and therefore also rock properties) may vary from place to place, from point to point or from one time to another.

Rock-slope failures may be classified according to the mode in which the applied stress exceeds the strength of the rock. Examples of slope failures are:

1. Rock fall,
2. Block flow,

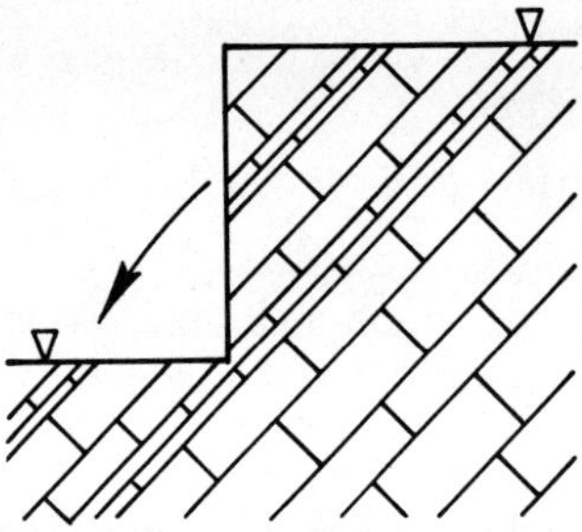

a) Unstable wall of rock: dip of rock strata toward excavation

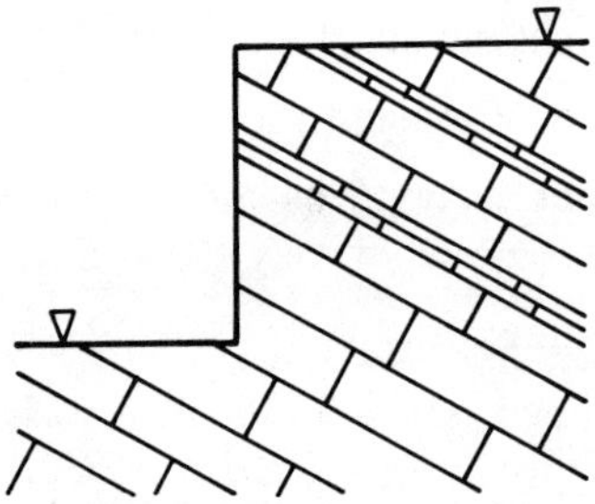

b) Stable wall of rock: dip of rock strata away from excavation

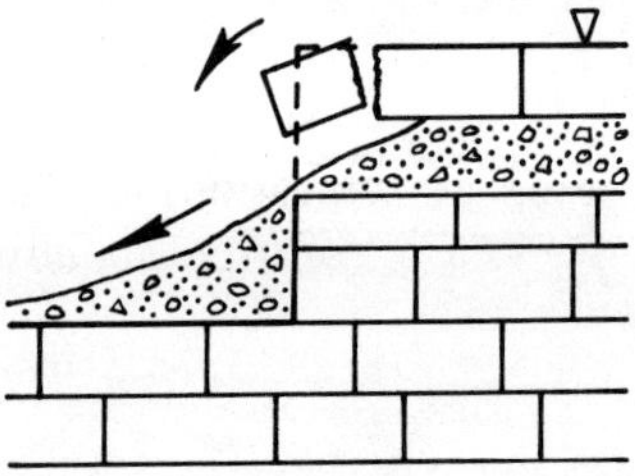

c) Unstable rock slope: undermining of overlying sound rock by loose fragments of shattered rock

Fig. 12-1
Unstable and stable excavation walls in rock

3. Slide along a pre-existing, dipped plane of discontinuity (bedding planes, faults, and joints, for example), and
4. Rotational slides.

Rock falls from cliffs and talus slopes have been studied in the field, and theoretically by rock trajectory evaluation. Detailed criteria for design of ditch sections along highways and of rock fences for protection against rock falls may be found in Reference 11. The criteria include fallout areas for energy dissipation, steep off-shoulder slopes to cope with angular momentum generated after impact, and rock fences acting as flexible buttresses.

Figures 12-1a, b, and c illustrate some unstable and stable rock slopes as a function of attitude of rock formations.

12-3. Stability of Rock Slopes

Almost all rock-slope failures occur along pre-existing, predetermined, natural rock discontinuities or planes of weakness, such as a fault or a shear zone, for example.

In rock-slope stability problems, the actual failure or sliding surface depends upon the spatial orientation, frequency, and distribution of the discontinuities, and the involved shear and interlock resistance to shear along them. Upon shearing, the geologic rock formation becomes unlocked, resulting in sliding (4-6, 12, 13).

Figure 12-2 shows a pre-existing geological discontinuity oriented at an angle of dip α with the horizontal. If the weight W of the rock mass superimposed upon the inclined discontinuity A-A_1 is large enough, and the

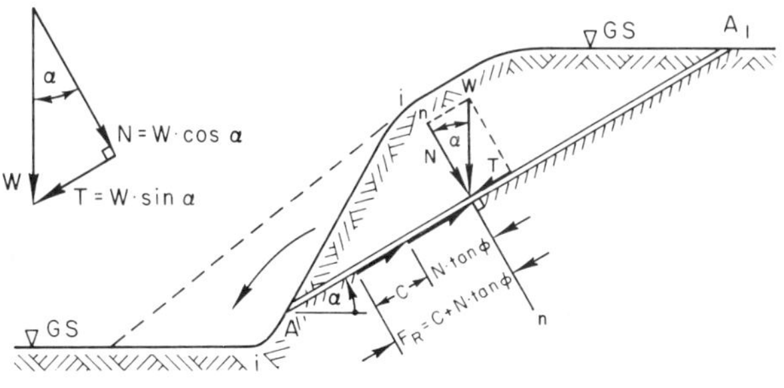

Fig. 12-2
Slope stability problem involving sliding over a geological discontinuity

resistance to shear F_R (or against sliding down of the rock mass) in this discontinuity is not sufficiently large enough, the superimposed rock mass above the discontinuity may become detached from the rock below the discontinuity and slide down over the discontinuity into the cut. Thus the man-made slope of the cut becomes ruined.

Basically, stability calculations of a rock slope over a pre-existing discontinuity such as a sliding surface is a problem of statics. For example, in Fig. 12-2, one determines the driving force F_D per unit length of run of slope

$$F_D = T = W \cdot \sin \alpha \tag{12-1}$$

causing shearing in the rock discontinuity A-A_1. Then, one computes the resisting force F_R to shear:

$$\begin{aligned} F_R &= N \cdot \tan \varphi + C \\ &= W \cdot \cos \alpha \cdot \tan \varphi + C \end{aligned} \tag{12-2}$$

where T = tangential (shear) force in the plane of discontinuity,
W = weight of the sliding rock mass over the shear plane,
α = angle of dip of slope,
φ = angle of friction,
$C = (c)\,(\overline{A - A_1})\,(1.0)$ = total cohesive (resisting) force along the shearing surface $(\overline{A - A_1})\,(1.0)$, and
$c = C/(\overline{A - A_1})\,(1.0)$ = cohesion per unit area.

The safety or stability of the rock slope in this example is expressed by means of the so-called factor of safety η against sliding (viz., shear) as a ratio of the resisting force F_R to the driving force F_D as

$$\eta = \frac{F_R}{F_D} = \frac{W \cdot \cos \alpha \cdot \tan \varphi + C}{W \cdot \sin \alpha} \tag{12-3}$$

When $C = 0$, then Eq. (12-3) simplifies to

$$\eta = \frac{W \cdot \cos \alpha \cdot \tan \varphi}{W \cdot \sin \alpha} = \frac{\tan \varphi}{\tan \alpha} \tag{12-4}$$

In such a case — for noncohesive material — the factor of safety is independent of the height of the slope and the form of the wedge.

If the total pore-water pressure U on the sliding surface is known, then

$$\eta = \frac{(W \cdot \cos \alpha - U) \cdot \tan \varphi}{W \cdot \sin \alpha} \tag{12-5}$$

If the factor of safety turns out to be unacceptable, the slope must be redesigned adopting a different angle of slope, or reinforced by means of rock anchors, or buttressed by means of a retaining wall. Provision for drainage facilities may decrease pore-water pressure in rock, as well as uplift forces.

12-4. Reinforcement of Rock Slopes

Rock-bolting can be used on natural or artificial slopes to prevent blocks of rock from falling away from the main mass when isolated by joints and faults. If the rock cut as shown in Fig. 12-2 is unstable ($\eta < 1$), then the rock slope can be stabilized by reinforcing it by means of a rock-anchoring system. The purpose of anchors is to transmit the stabilizing force into the rock slope comprehending the rock joints, viz., discontinuities. By means of

a) Rock-anchored vertical slope

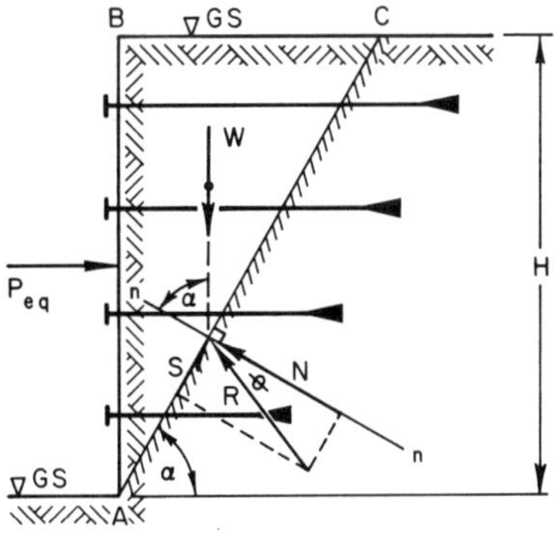

b) Force triangle-solving for P_{eq}

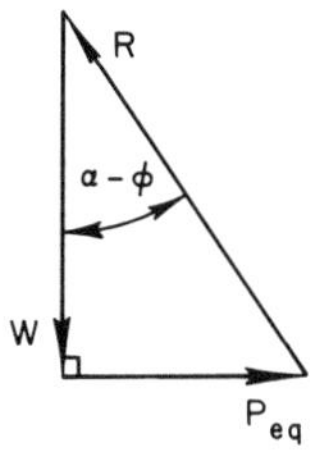

Fig. 12-3
Rock-anchor system

the installed rock anchorage, the joints close more or less. Thus the friction between the surfaces of separation in the anchorage zone is increased.

Now, for the purpose of illustrating determination of the necessary anchor force, refer to Figs. 12-3a and b. Assume an unstable, vertical slope. To maintain force equilibrium of this slope, a horizontal rock-anchor system is introduced. These anchors supply the necessary force P_{eq} on the rock for equilibrium. The magnitude of this necessary force P_{eq} for equilibrium can be conveniently solved graphically by means of the force triangle, as shown in Fig. 12-3b.

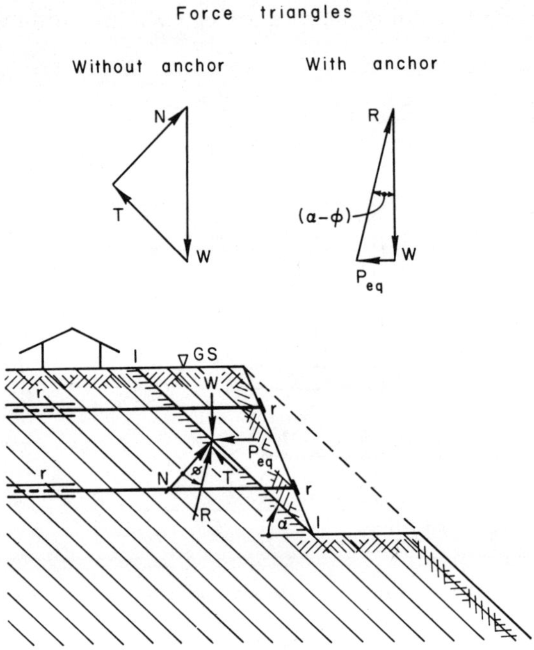

Fig. 12-4
Anchoring of a rock slope
$l-l$: Potential failure line
$r-r$: rock anchor
W: weight of ruptured rock wedge above plane 1—1
φ: angle of friction along plane 1—1
N: normal force
T: tangential shear resistance
R: reaction
P_{eq}: anchor force

The actual total rock-anchor force P is obtained by multiplying the equilibrium force P_{eq} by a prescribed factor of safety η:

$$P = \eta \cdot P_{eq} \tag{12-6}$$

For a certain number of chosen anchors, the force P determines their diameter as well as their spacing in the face of the slope. The length of the anchors is that which is necessary to cross the discontinuity plus the needed fixing (keying) or bonding length of the anchor in the rock. Thus the rock anchors would prevent massive falls of loose rock fragments from the steeply cut, reinforced rock face.

High slopes may be constructed by parts, separating the upper and lower slopes by a berm (Fig. 12-4).

Unstable rock slopes may be kept stable, also, by means of retaining walls. Figure 12-5 shows a rock slope stabilized by means of a retaining wall and several rock anchors.

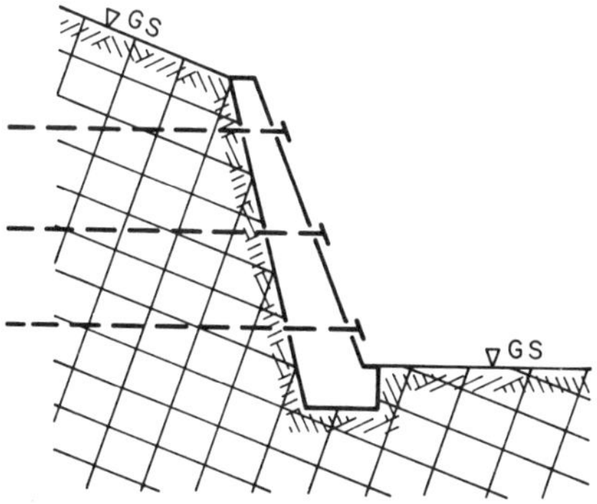

Fig. 12-5
Anchoring of a rock slope with several anchors, and stabilization by means of a retaining wall

If the retaining wall is founded on a firm rock, a single anchor may be used, depending upon the height of the wall. By means of anchoring, a considerable reduction in thickness of the retaining wall may be achieved. The rock slopes may also be reinforced by means of anchors directly, without constructing any retaining walls.

Sometimes fissured rock may also be stabilized by means of cement grouting.

Depending upon the geologic nature and jointing pattern of the rock, in evaluating the stability of rock slopes each individual rock stabilization project requires an individual approach, analysis, and solution (11-13).

12-5. Effect of Precipitation on Stability of Rock Slopes

The permeability of rock to water depends upon the presence of open and continuous fissures and seams within the rock mass. Part of the precipitation water intrudes through the surface cracks into the rock. Depending upon the intensity of precipitation, this water may bring about a high or low water table.

Generally, the difference in water table in the ground during dry and wet seasons may vary by several meters. However, in rock, the position of the water table is indeterminate because of the various kinds, frequency, and distribution of seams, cracks, and other kinds of voids in the rock.

During intensive rainfall, all of the rainwater cannot be absorbed by the rock. Most of the water runs down along the sloping ground surface.

Also, during the maximum spring thaw of snow, most of the meltwater usually runs down, because the cracks near the ground surface of the slope are still frozen. Also, the meltwater of the ice from within the fissured rock cannot escape. Frequently, these conditions, too, impair adversely the stability of slopes in fissured rock, bringing about rock bursts and rock slides.

12-6. Rock-Slope Stability Based on Rock Strength

Nowadays, the most frequently used theory for evaluating, theoretically, the strength and stability of a homogeneous isotropic material is the one put forward by MOHR (9). According to MOHR's theory, rupture of a material takes place along surfaces where the tangential, viz., shear, stresses attain a definite value. At limit equilibrium, by COULOMB's shear strength equation, the magnitude of the shear stress τ is a function of the effective normal stress, namely:

$$\tau = f(\sigma_{n_{eff}}) = \sigma_{n_{eff}} \cdot \tan \varphi + c \tag{12-7}$$

where $\sigma_{n_{eff}}$ = effective normal stress on the rupture plane (Fig. 12-6),

$\tan \varphi$ = coefficient of friction, and

c = test parameter known as the cohesion.

In practice, the limiting stability or equilibrium condition of a homogeneous rock element can be conveniently expressed with a satisfactory degree of precision by means of principal stresses σ_1 and σ_3 in the form of Coulomb's shear strength formula, Eq. (12-7).

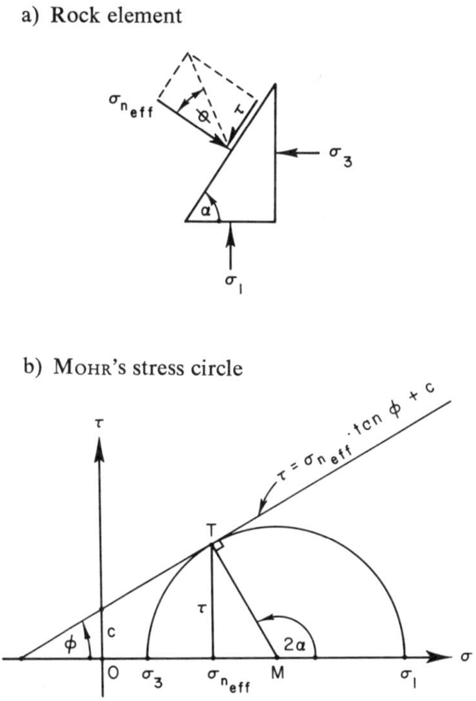

Fig. 12-6
Coulomb-Mohr strength condition

From Mohr's stress circle, Fig. 12-6, obtain:

$$\tau = \frac{\sigma_1 - \sigma_3}{2} \cdot \cos\alpha = \left(\frac{\sigma_1 + \sigma_3}{2} - \frac{\sigma_1 - \sigma_3}{2} \cdot \sin\varphi\right) \cdot \tan\varphi + c \qquad (12\text{-}8)$$

After rearrangement, the major principal stress σ_1 is written as

$$\sigma_1 = \sigma_3 \left(\frac{1 + \sin\varphi}{1 - \sin\varphi}\right) + \frac{2 \cdot c \cdot \cos\varphi}{1 - \sin\varphi} \qquad (12\text{-}9)$$

or

$$\sigma_1 = \sigma_3 \cdot K_p + 2 \cdot c \cdot \sqrt{K_p} \qquad (12\text{-}10)$$

or

$$\sigma_1 = \sigma_3 \cdot \tan^2\left(45^\circ + \frac{\varphi}{2}\right) + 2 \cdot c \cdot \tan\left(45^\circ + \frac{\varphi}{2}\right) \tag{12-11}$$

where

$$(1 + \sin\varphi)/(1 - \sin\varphi) = K_p \tag{12-12}$$

and

$$\cos\varphi/(1 - \sin\varphi) = \sqrt{K_p} \tag{12-13}$$

K_p is the COULOMB's passive earth pressure (or earth resistance) coefficient.

The degree of stability of a homogeneous rock slope at any point of the slope is usually expressed by means of a factor of safety η if the principal stresses and rock strength parameters φ and c are known. The factor of safety η is then expressed as the ratio of the rock strength,

$$\sigma_R = \sigma_1 = \sigma_3 \cdot K_p + 2 \cdot c\sqrt{K_p} \tag{12-14}$$

where σ_R = resisting stress at failure to the acting major stress, and

$$\sigma_D = \sigma_{1\,\text{acting}} \tag{12-15}$$

where σ_D = driving stress:

$$\eta = \frac{\sigma_R}{\sigma_D} = \frac{\sigma_3 \cdot K_p + 2 \cdot c \cdot \sqrt{K_p}}{\sigma_{1\,\text{acting}}} \tag{12-16}$$

the assumption herein being that the rock strength parameters φ and c do not depend upon the stress condition of the rock massif, nor upon the depth of the point under consideration in the rock below the ground surface.

When $\sigma_3 = 0$, then

$$\tau_{max} = \sigma_1/2 \tag{12-17}$$

and

$$\sigma_1 = 2 \cdot \tau_{max} \tag{12-18}$$

Substitution of these values into the η-equation, Eq. (12-16), renders the factor of safety η against rupture of a rock slope in its lower part as

$$\eta = \frac{c \cdot \sqrt{K_p}}{\tau_{max}} = \frac{c \cdot \tan \cdot \left(45^\circ + \frac{\varphi}{2}\right)}{\tau_{max}} \tag{12-19}$$

Thus, for an unconfined condition ($\sigma_3 = 0$), the factor of safety η of a rock slope is a function of the maximum shear stress τ_{max}.

By means of the η-equation, it is possible to calculate factors of safety for a number of assumed points in the slope, i.e., $\eta = f(\alpha_a)$, where α_a is the angle between the horizontal and a line drawn from the toe-point of the slope to the assumed point in the slope. Connecting points of the same factor of safety by means of a curve renders a field of isocurves, viz., zones of relative strength of the slope of the rock.

12-7. Closing Remarks about Stability of Rock Slopes

Evaluation of stability of rock slopes may be made if the nature, magnitude, and distribution of natural, internal stress conditions in the slope-forming rock massif are known. Unfortunately, in rock engineering, up to now, the problem of determining natural stress conditions in rock has been worked out very little methodologically. The solution to this problem requires a great deal of theoretical as well as experimental research. One should rcognize that the factor "stressed condition" in rock is one of the

Fig. 12-7
Concentration of stresses in the lower part of a slope
(Redrawn from Hoek and Gralewska, 1969)

determining factors in the behavior of rock slopes, and thus in the degree of stability of slopes.

Research has shown that, basically, there occur large, internal stress concentrations in the lower part of the slope in the neighborhood around and below the toe of the slope as shown qualitatively in Fig. 12-7, probably because of the great weight of the rock near the slope toe, and because of residual tectonic stresses in the rock. The large stress concentrations in this part of the slope may be the source of rock bursts and other kinds of rock failure.

Open-pit mining engineering practice has shown that rock in the lower part of the slope of a quarry fails in shear brought about by tangential or shear stresses. Therefore, in studying stressed conditions of rock slopes, major attention should be devoted to determining the magnitude of the maximum shear stresses in the lower part of the slope of the rock.

REFERENCES

1. Cording, E. J., 1968, "Stability of Large Underground Openings at the Nevada Test Site," *ASCE Seminar on Rock Mechanics in Civil Engineering Practice,* held April 9–10 and May 7–8, 1968, in New York, N. Y., ASCE Metropolitan Section, Foundations and Soil Mechanics Group.
2. Cording, E. J., 1976, "Observations and Design in the Construction of Rock Chambers in Urban Areas," *ASCE Rock Excavation Seminar,* held October 1976 in New York, N. Y., ASCE Metropolitan Section, Foundations and Soil Mechanics Group.
3. Farmer, I. W., 1968, *Engineering Properties of Rocks.* London: E & F. N. Spon Ltd.
4. Highway Research Board, 1963, *Stability of Rock Slopes.* Washington, D. C.: *Highway Research Record No. 17,* National Research Council (52 pages).
5. Hoek, E., and J. W. Bray, 1977, Revised 2nd edition, *Rock Slope Engineering.* London: The Institution of Mining and Metallurgy.
6. Jaeger, J. C., 1971, "Friction of Rocks and Stability of Rock Slopes," Eleventh Rankine Lecture, *Géotechnique* (London), Vol. 21, No. 2, pp. 97–134.
7. Jumikis, A. R., 1956, "Rupture Surface in Sand under Oblique Loads," *Proceedings ASCE, Soil Mechanics and Foundations Division,* Vol. 82, Paper 861, pp. SM 1 to SM 6.
8. Lang, T., 1962, "Theory and Practice of Rock Bolting," *Transactions American Institute of Mining and Metallurgical Engineers, Mining Division,* Vol. 223.
9. Mohr, O., 1900, "Welche Umstände bedingen die Elastizitätsgrenze und den Bruch eines Materials?", in *Zeitschrift des Vereins Deutscher Ingenieure,* Vol. 44, p. 1524.
10. Patton, F. D., 1968, Lecture notes on "The Determination of Shear Strength of Rock Masses," Lecture given on May 7, 1968 at the *ASCE Seminar on Rock Mechanics in Civil Engineering Practice* to the ASCE Metropolitan Section, Foundations and Soil Mechanics Group, in New York City, p. 12.

11. RITCHIE, A. M., 1963, "Evaluation of Rockfall and its Control," Washington, D. C.: *Highway Research Record No. 17,* Highway Research Board, May 1963, pp. 13–28.

12. ZARUBA, Q., and V. MENZL, 1969, *Landslides and Their Control.* Amsterdam: Elsevier.

13. ZISCHINSKY, U., 1966, "On the Deformation of High Slopes," *Proceedings First Congress of the International Society for Rock Mechanics,* Vol. 2, p. 179.

OTHER RELATED REFERENCES

ASCE – *Rock Engineering for Foundations and Slopes.* New York, N. Y.: Vol. 1, 1976; Vol. 2, 1977.

HOEK, E., and A. GRALEWSKA, *A Rock Mechanics Information Service.* London: Imperial College Rock Mechanics Research Report No. 9, May 1969.

JOHN, K. W., 1968, "Graphical Stability Analysis of Slopes in Jointed Rock," *Proceedings ASCE, Soil Mechanics and Foundations Division,* Vol. 94, SM 2, pp. 497–526.

KOMARNITSKII, N. I., 1968, *Zones and Planes of Weakness in Rocks and Slope Stability.* New York, N. Y.: Consultants Bureau.

LADANYI, B., and G. ARCHAMBAULT, 1970, "Simulation of Shear Behavior of a Jointed Rock Mass," *Proceedings 11th Symposium on Rock Mechanics,* New York, pp. 105–125.

PETZNY, H., 1967, "Über die Stabilität von Felshängen," in *Felsbau in Theorie und Praxis – Rock Engineering in Theory and Practice,* L. MÜLLER and C. FAIRHURST (editors). Wien: Springer-Verlag, pp. 37–57.

WITTKE, W. et al., 1972, "Three Dimensional Calculation of the Stability of Caverns, Tunnels, Slopes and Foundations in Anisotropic Jointed Rock by Means of the Finite Element Method," in *Deutsche Beiträge zur Geotechnik.* Essen: Deutsche Gesellschaft für Erd- und Grundbau, Vol. 1, No. 1.

WOOD, A. M., and N. E. V. VINER-BRADY, 1955, "Folkstone Warren Landslip Investigation, 1948–1950;" "Folkstone Warren Landslip Remedial Measures, 1948–1950," Proceedings, Institute of Civil Engineers, London, Vol. 4.

APPENDICES

APPENDIX 1

GREEK ALPHABET

Α	α	Alpha	Ν	ν	Nu
Β	β	Beta	Ξ	ξ	Xi
Γ	γ	Gamma	Ο	o	Omicron
Δ	δ	Delta	Π	π	Pi
Ε	ε	Epsilon	Ρ	ϱ	Rho
Ζ	ζ	Zeta	Σ	σ ς	Sigma
Η	η	Eta	Τ	τ	Tau
Θ	ϑ	Theta	Υ	υ	Upsilon
Ι	ι	Iota	Φ	φ	Phi
Κ	$\varkappa$	Kappa	Χ	χ	Chi
Λ	λ	Lambda	Ψ	ψ	Psi
Μ	μ	Mu	Ω	ω	Omega

APPENDIX 2

KEY TO SIGNS AND NOTATIONS

Symbol	
A	Area; circumferential shear area; constants; rock element
A_0	Initial, unnecked, cross-sectional area of a rock test specimen
ASCE	American Society of Civil Engineers
ASTM	American Society for Testing and Materials
a	A coefficient; length; a ratio: $a = r_0/r_i$; cross-sectional area of anchorage; *are* (1 *are* = 1 a = 100 m^2)
B	A constant
b	A coefficient; width of gauge section; length
C	Coefficient; compression; total amount of cohesion
Cal	Large or kilo-calorie = 1000 gram-calories
C-fl	Point of counter-flexure
℄	Center line
c	Cohesion; a coefficient; test parameter; distance
cal	Small calorie
c_{eff}	Cohesion on effective stress basis
c_m	Mass heat capacity
c_{mi}	Mass heat capacity of ice
c_{mr}	Mass heat capacity of rock and dry soil mineral particles
c_{mw}	Mass heat capacity of water
const	A constant
c_v	Volumetric heat capacity
c_{vf}	Volumetric heat capacity of frozen soil or rock

c_{vu}	Volumetric heat capacity of unfrozen soil or rock
D	Diameter; width of opening; width of footing
d	Diameter
dr	Differential radius
ds	Differential arc
dT	Differential temperature
du	Differential displacement
dV	Differential volume
$d_{\omega=0}$	$= r - r_i =$ thickness of tensile zone above the crest
Δd	Diametral deformation
Δdr	Change in differential radius
$d\sigma_r$	Differential radial stress
$d\omega$	Differential angle
E	Modulus of elasticity
E_{ave}	Average modulus of elasticity
E_c	Compression modulus or modulus of deformation; modulus of elasticity of a confined rock
E_i	Initial tangent modulus at zero load
E_{rock}	Modulus of elasticity of rock
E_s	Secant modulus of elasticity for a particular point
E_t	Tangent modulus of elasticity on the stress-strain curve at a particular point
$E_{t\,50}$	50 %-tangent modulus
e	Void ratio; base of the system of natural logarithms
e-e	An envelope line
e_{el}	$= \varepsilon_{el}/\varepsilon_{total} =$ ratio of elastic strain to total strain
e_{pl}	$= \varepsilon_{ir}/\varepsilon_{total} =$ ratio of plasticity strain to total strain
F	Force; frictional force; resistance to shear force
F_D	Driving force
F_R	Resistance to shear; resisting force
f	A function of . . .; coefficient of static friction: $f = \tan \varphi$
f'	Coefficient of dynamic friction
G	Specific gravity
G_b	Bulk or apparent specific gravity
G.S.	Ground surface
GW	Groundwater
GWT	Groundwater table
g	Acceleration of gravity; gram

H	Symbol for rating of hardness of minerals; horizontal force
HRB	Highway Research Board
h	Hour; pressure head; height; thickness
Δh	Deformation; compression; change in height
I	Moment of inertia (axial)
I_T	Toughness index of rock
ISRM	International Society of Rock Mechanics
i	Hydraulic gradient; thermal gradient; pressure gradient; angle of slope of an inclined shear plane; imbrication
i_{cr}	Critical hydraulic gradient
K	Coefficient of thermal conductivity
K_a	COULOMB's active earth pressure coefficient
K_p	COULOMB's passive earth pressure coefficient
k	Kilo; permeability to water; kip (1 k = 1000 lb)
kg	Mass kilogram
kg_f	Force kilogram
kg/m^3	Mass kilogram per cubic meter
kg_f/cm^2	Force kilogram per square centimeter
kN/m^2	Kilo newtons per square meter
kN/m^3	Kilo newtons per cubic meter
L	Length, distance; latent heat of fusion = 80 cal/g
L_t	New length upon thermal expansion
ΔL	Change in length; linear deformation; displacement
dL	Differential length
l	Liter
l/s	Liters per second
M	Mega; bending moment
M_R	Modulus ratio
MN/m^2	Mega newtons per square meter
MN/m^3	Mega newtons per cubic meter
m	Meter; POISSON's number: m $1/\mu$; displacement coefficient; mass
max	Maximum
min	Minimum
m_r	POISSON's number of rock
N	Newton; normal load
N/m	Newtons per square meter

n	Porosity; relative volume of voids; number of specimens; number of bolts; function of support rigidity and strength of rock mass
n_a	Relative volume of air
n_m	Relative volume of water
n_s	Relative volume of solids
n_w	Relative volume of water
n'	Function of rock mass quality and the kind of support method
P	Load; force; ultimate punching load; total rock anchor force
P_N	Normal pressure
P. L.	Proportional limit; plastic limit
P_{eq}	Force on rock for equilibrium; anchor force
p	Pressure
p_h	Horizontal stress
p_i	Initial stress $= c \cdot \cot \varphi$; inside pressure; bolt pressure on rock
p_{max}	Maximum pressure
p_{min}	Minimum pressure
p_o	Outside pressure
psi	Pounds per square inch
p_v	Vertical stress
Q	Discharge; amount of heat flow; heat energy
Q_L	Total latent heat of fusion
R	Radius; resultant force; reaction
r	Radius; polar coordinate
r_i	Initial radius; inside radius
r_o	Outside radius
Δr	Change in radius
S	Degree of saturation; total shear resistance
S_f	Softening factor
S_t	Dry/wet factor
s	Second; shear strength
s_{ave}	Average punching shear strength
s_1, s_2, s_3, s_4	Shear strength
T	Temperature; horizontal shear force; tangential force; torque; tension

$\overline{T}$	Effective shear force
T_f	Freezing temperature
T_i	Initial temperature
T_o	Initial temperature
T_s	Surface temperature
ΔT	Temperature increment; temperature decrement; temperature difference; rise in temperature; change in temperature
T_1	Final temperature
t	Symbol for a metric force ton; time; thickness of gauge section; thickness of specimen
t_f	Metric force ton
t_f/m^3	Metric force ton per cubic meter
$\tan \varphi$	Coefficient of friction
t-t	A tangent line
U	Neutral force; uplift pressure; hydrostatic pressure; groundwater pressure
u	Neutral stress; pore water pressure; displacement; radial displacement; deformation
u_i	Inside displacement
u_o	Outside displacement
V	Volume; total volume
V_a	Volume of air or gas
V_s	Volume of solids
V_t	Total volume; new volume after thermal expansion
V_v	Volume of voids
V_w	Volume of water
ΔV	Volume element
v	Velocity; index for vertical
v_{cr}	Critical velocity
W	Weight; total weight; weight of loosened rock; weight of ruptured rock wedge
W_{air}	Weight in air
W_d	Dry weight
W_s	Weight of solids
W_{sat}	Weight of saturated soil (rock)
$\mathrm{W}_{sat\ in\ air}$	Weight of saturated soil (rock) in the air
$W_{sat\ in\ water}$	Weight of saturated soil (rock) in water
W_w	Weight of water
w	Natural moisture content; displacement; deformation
w_o	Surface displacement of rock

w_{sat}	Saturation moisture content
X	Force-field component
x	An unknown; length; a coordinate
Y	Yield point; force-field component
y	Deflection; length; an unknown; a coordinate
Z	Section modulus $= I/c$; force-field component
z	An unknown; depth; a coordinate
α	Alpha. Angle; a constant; a ratio: $\alpha = r_0/r$; coefficient of thermal diffusivity or temperature conductivity; angle of rupture; angle of dip; angle of slope
α_a	Angle
α_t	Coefficient of linear thermal expansion
β	Beta. Coefficient of volumetric thermal expansion; a constant; angle
γ	Gamma. Unit weight
γ_d	Dry unit weight
γ_s	Unit weight of solids
γ_{sat}	Saturated unit weight
γ_{sub}	Submerged unit weight
γ_w	Unit weight of water
Δ	Delta. Difference; increment; decrement
ΔH, ΔL	Change in length; deformation
ΔT	Change in temperature
Δd	Change in diameter; diametrical deformation
Δdr	Change in elementary radius dr
Δh	Absolute deformation
Δp	Pressure increment
Δr	Absolute radial deformation
Δr_d	Thickness of zone of blasting disturbance
Δt	Time increment
Δu	Absolute deformation

Δx, Δy, Δz	Coordinate increments
$\Delta\varepsilon$	Change in strain
$\Delta\sigma$	Change in stress; standard deviation
δ	Delta. Angle; deformation
ε	Epsilon. Strain; relative deformation; angle
ε_{el}	Elastic strain
ε_{h}	Horizontal strain
ε_{ir}	Irreversible, permanent strain
ε_{pl}	$= \varepsilon_{ir} =$ permanent strain
ε_{t}	Tangential strain component
ε_{total}	Total strain
ε_{v}	Vertical strain
ε_{x}	Lateral strain
ε_{y}, ε_{z}	Longitudinal strains
ε_{1}	Longitudinal strain; bending tensile strain
ε_{2}	Bending compression strain
ε_{3}	Lateral or transverse strain
ζ	Zeta. A coefficient; zeta potential
η	Eta. Factor of safety
θ	Theta. Angle.
λ	Lambda. A coefficient; ratio
λ_{o}	A constant
μ	Mu. POISSON's ratio: $\mu = 1/m$; coefficient of friction
ξ	Xi. A coefficient
ϱ	Rho. Radius; angle; density
Σ	Sigma. Summation of . . .
ΣF_{r}	The sum of projected forces in radial direction
σ	Sigma. Stress; normal stress; compressive stress
σ_{A}, σ_{B}, σ_{D}	Stresses at point A, B, and D, respectively
σ_{D}	Driving stress
$\sigma_{P.L.}$	Stress at plastic limit
σ_{R}	Resisting stress at failure
σ_{Y}	Yield stress
σ_{ave}	Average stress; average strength of rock specimen
σ_{c}	Compressive stress
σ_{cu}	Ultimate compressive strength of an equivalent cubical rock specimen whose ratio $b/h = 1.0$

σ_{eff}	Effective stress
σ_h	Horizontal stress
σ_{max}	Maximum stress
σ_{min}	Minimum stress
σ_n	Normal stress
$\bar{\sigma}_n$	Effective normal stress
$\sigma_{n_{eff}}$	Effective normal stress
$\sigma_{n1}, \sigma_{n2}, \sigma_{n3}, \sigma_{n4},$	Normal stresses
σ_o	Contact pressure on rock (soil)
σ_{pr}	Precompression stress
σ_r	Radial stress
$\sigma_{re\,1}$	Radial elastic primary stress
$\sigma_{re\,2}$	Radial elastic secondary stress
$\sigma_{r\,el}$	Radial stress in the elastic zone
$\sigma_{r\,max}$	Maximum radial stress
σ_{ro}	Radial stress for the outer contour
σ_{rp}	Radial stress in the plastic zone
σ_t	Tensile stress; tangential stress
$\sigma_{te\,2}$	Secondary tangential elastic stress
$\sigma_{t\,el}$	Tangential stress in the elastic zone
$\sigma_{t\,max}$	Maximum normal tangential stress
$\sigma_{t\,min}$	Minimum normal tangential stress
σ_{to}	Tangential stress for the outer contour
σ_{tp}	Tangential stress in the plastic zone
$\sigma_{t_{ult}}$	Uniaxial ultimate tensile strength
σ_u	Unconfined compressive strength
σ_{ult}	Ultimate stress
σ_{unc}	Unconfined compressive strength
σ_v	Vertical stress; primary stress
$\sigma_x, \sigma_y, \sigma_z$	Stress components
σ_1	Major principal stress
$\sigma_1 - \sigma_3$	Deviator stress
σ_2	Intermediate principal stress
σ_3	Minor principal stress
$\sigma_{1_{eff}}$, $\sigma_{3_{eff}}$	Effective principal stresses
τ	Tau. Shear stress; shear strength;
$\bar{\tau}$	Effective shear stress
τ_e	Total elastic shear stress in elastic-plastic boundary

τ_{e1}	Primary elastic shear stress
τ_{e2}	Secondary elastic shear stress
τ_{el}	Shear stress in elastic zone
τ_{max}	Maximum shear stress
τ_{p}	Shear stress in plastic zone
τ_{rt}, τ_{tr}	Shear stresses in polar coordinate system
τ_{xy}, τ_{yz}, τ_{zx}	Shear stresses
Φ	Phi. AIRY's stress function
φ	Angle of friction; test parameter
φ_{eff}	Angle of friction on effective stress basis
φ_{r}	Angle of residual shear resistance
φ_{u}	Test parameter
φ_{unc}	Angle of friction obtained from unconfined compression test
ω	Omega. Angle; amplitude; a polar coordinate

APPENDIX 3

GLOSSARY OF TERMS

Abutment The place of abutting. The part of a structure that directly receives thrust or pressure (as of a supporting zone of an arch, vault, end-pier of a bridge abutted against soil).

Adit A nearly horizontal passage from the surface in a mine.

Anisotropy A condition of a material having different properties in different directions. For example: the state of geologic strata of transmitting sound waves with different velocities in the vertical and in the horizontal directions. Situation of a material having different moduli of deformation in different directions.

Anticline An arch-like fold (up-fold) in rocks, with the beds dipping in opposite directions from the crest.

Arch The configuration of the upper part of a tunnel section above the springline. The crown. The curved roof of an underground opening (tunnel).

Are Area (1 are = 1 a = 100 m^2)

Attitude Position with respect to the direction of the compass and to the horizontal.

Basalt An extrusive rock. Generally, any fine-grained, dark-colored igneous rock.

Bedding plane Plane dividing sedimentary rocks of the same or

	different lithology. The division planes which separate the individual layers, beds, or strata marking the boundary between a bed and the bed above and/or below it.
Bedrock	Any layer of rock underlying soils. Geologically, the term denotes material underlying drift deposits.
Bending strength	An alternate term for flexural strength.
Bond strength	Stress required to rupture a bond in a material, or two kinds of material cemented one to another (a concrete block cemented on rock).
Break	Separation into parts with suddenness or violence. A burst. Fragmentation of solid rock as a result of blast.
Breakdown of structure	1) Occurring often quite gradually under an increasing load. 2) Displacements of crystal grains accompanied by a partial destruction of the cohesion, occurring under an increasing load. This results in a gradual loosening or tearing apart of the structure of the material under increasing stress.
Breaking load	Load which causes fracture in compression, tension, flexure or torsion tests.
Breccia	Sedimentary rock consisting largely of angular fragments embedded in a fine-grained matrix.
Brittle	A material is said to be in a brittle state or brittle under conditions in which its stability to resist load decreases with increasing deformation.
Brittle fracture	Failure or rupture of a material with little or no plastic flow of deformation of a rock crystal lattice. Usually this type of failure is associated with impact loads. However, many materials at low temperatures also show brittle fracture failures under static loads.
Brittleness	A property of material that ruptures or fractures with little or no plastic flow.
Burst	To break suddenly into pieces from impact or from pressure within.
Calcareous	Containing calcium carbonate.
Calcite	A mineral, calcium carbonate ($CaCO_3$). The principal constituent of limestone.
Calcium carbonate	A solid, $CaCO_3$, occurring in nature (calcite, for example).

Carbonate A compound containing the radical CO_3—.

Cavern A subterranean hollow; an underground cavity. Most freqent in limestones and dolomites.

Cohesion The property of rock (soil) particles to bind together.

Competent ground Ground that does not require support when an underground opening (tunnel) is excavated through it.

Competent rock A rock which is capable of sustaining an underground opening or a steep slope at the earth's surface without artificial support. Also, a rock which is sufficiently strong to transmit forces under given conditions. Rocks which are sufficiently plastic to deform without fracturing are incompetent.

Conglomerate Sedimentary rock consisting largely of rounded pebbles cemented together by another mineral substance.

Consolidation Geologically, any or all of the processes whereby loose, soft, or liquid earth materials become firm and coherent.

In soil mechanics, the adjustment of a saturated soil in response to increased static load. The process involves the squeezing of water from the pores of the soil and a decrease in its void ratio in time.

Continuity Something absolutely continuous and homogeneous of which no distinction of content can be affirmed except by reference to something else (as duration). An uninterrupted, ordered sequence. An identity of substance-uniting discrete parts.

Continuum Continuum is a mathematical abstraction applied to a large collection of material particles.

Creep Time-dependent deformation or strain. Deformation that occurs over a period of time when a material is subjected to constant stress at constant temperature. Slow deformation that results from long application of stress.

An imperceptibly slow, more or less continuous, downward and outward movement of slope-forming rock or soil.

Creep limit Alternate term for creep strength.

Creep strength Maximum stress required to bring about a specified amount of creep in a specified time.

Critical stress Maximum and minimum compressive stresses on the boundary of an opening.

Diastrophism All movements of the solid part of the earth, resulting in its displacement (faulting) or deformation (folding).

Dilatancy Refers to a relative increase in volume. Packed sand expands when sheared.

Dilation Deformation, that is, change in volume but not in shape.

Dip The slope of a bed of rock relative to the horizontal.

Ductility Extent to which a material can sustain plastic deformation without breaking or rupture. Elongation and reduction of area are common indices of ductility.

Elastic Capable of sustaining stress without permanent deformation.

Elasticity The property or quality of being elastic.

Elastic limit The least stress without undergoing permanent deformation. Also called the yield point.

Energy Ability to do work.

Erosion Processes whereby earthy or rock material is loosened or dissolved and removed from any part of the earth's surface. It includes weathering, solution, and transportation.

Erosion agents: running water, waves, moving ice, and wind.

Extrusive (extrusion) This term is applied to igneous rocks which emerge at the surface. Those igneous rocks were derived from magmas or magmatic materials poured out or ejected at the earth's surface, as distinct from the intrusive or plutonic igneous rocks that have solidified from magmas that have been injected into older rocks at depth without reaching the surface. (Effusive rocks; volcanic rocks).

Failure Failing to perform an expected action. In a general sense, failure includes both fracture and flow.

Fatigue Permanent structural change that occurs in a material subjected to fluctuating stress and strain. In general, fatigue failure can occur with stress levels below the elasticity limit.

Fault A fracture in the rock of any dimension.

Flow Unrestricted plastic deformation.

Fold A bend in a rock of any form.

Force field — A region of space or a medium where at every point at any time there acts a definite, coordinated force — the so-called force field.

Formation — A group of rocks with recognizable and traceable boundaries, sufficiently alike lithologically to be mapped as a unit.

Fracture — A general term for discontinuities. A surface of discontinuity formed by rock failure under stress. Rock cracked or broken into fragments along planes other than joints or bedding. Breaks in rocks caused by intense faulting and folding. Fracture includes faults, joints, cracks, and other breaks.
The term fracture is used in the sense of brittle fracture to imply a complete loss of cohesion across a surface. Failure by parting of the material. Distinct surfaces of separation in the material.

Gallery — A subsidiary passage in a cave at a higher level than the main passage.
A horizontal or nearly horizontal underground passage either natural or artificial.

Geology — The study of the earth.

Geotechnics (géotechnique) — The discipline of foundation engineering. In its modern form this term is synonymous with "Soil Mechanics" or "Earthworks" (such as roads, tunnels, foundations, dams) in any way associated with engineering in rock and soil.

Geotectonic — Pertaining to the form, arrangement, and structure of the rock masses composing the earth's crust.

Geothermic (geothermal) — Pertaining to the heat of the earth's interior.

Geothermic degree — The average depth within the earth's crust corresponding to an increase of one degree in temperature.

Geothermal gradient — The change in temperature of the earth with depth, expressed either in degrees per unit depth or in units of depth per degree.

Geyser — Literally, a roarer; intermittent hot springs or fountains. Columns of water are thrown out at intervals with great force.

Gneiss — A coarsely banded metamorphic rock.

Gouge — Finely abraded material occurring between the walls of a fault, the result of grinding movement.

	Filling material such as silt, clay, rock flour and other kind of geological debris in joints, cracks, fissures, faults, and other discontinuities in rock.
Granite	A coarse-grained, igneous rock with abundant quartz and feldspar.
Ground	The material through which a tunnel is driven, whether it is a solid rock, unconsolidated rock, or running mud.
Ground arch	The rock located immediately above a tunnel which transfers the overburden load onto rock located on both sides of the tunnel.
Groundwater	That part of the subsurface water that has collected to form an underground "reservoir" as determined by the porosity and degree of fissuring of rocks and by the underlying impervious strata. The water-logged strata below the groundwater surface is the zone of saturation. The groundwater surface itself is the groundwater table.
Grout	Thin mortar.
Heterogeneous	A characteristic of a medium or a field of force that signifies that the medium has properties that vary with position within. Having unlike qualities. Opposed to homogeneous.
Homogeneous	Of the same nature, consisting of similar parts or elements of a like character. Having the same properties at all points. Uniform and continuous in all parts. A uniform composition throughout, as opposed to heterogeneous.
Hooke's law	Strain is proportional to stress.
Hoop stress	Circumferential stress in a material of cylindrical form subjected to internal or external pressure.
Igneous rocks	Strictly speaking, rocks which have been molten at some time in their history.
Impact	Collison of two masses.
Impact strength	The ability of a material to withstand shock loading. The work done in fracturing a test specimen in a specified manner under shock loading.
Incompetent rock	Rock incapable of standing in underground openings or steep slopes at the surface without support.
In situ	In its natural position or place.
Intact rock	A material which can be sampled and tested in the

laboratory, and which is free of large-scale structural features such as joints, bedding planes, shear zones, and other rock defects.

Internal friction Forces within a rock or soil which offer resistance to externally applied shear forces.

Intrusive (intrusion) Applied to igneous rocks which have been emplaced below the surface. Igneous rocks that invade older rocks.

Intrusive rock Rock that consolidated from magma beneath the surface of the earth.

Isotropic Having the same properties in all directions (said of a medium with respect to elasticity, conduction of heat or electricity, or radiation of heat or light).

Landslide The perceptibly downward and outward movement of slope-forming materials such as rock, soil, artificial fill, or a combination of these materials. Also, earth and rock that become loosened from a hillside by water or snow or an earthquake, and slide or fall down the slope.

Lava Molten volcanic rock extruded at the surface.

Limestone A sedimentary rock composed largely of calcium carbonate.

Lithology The physical character of a rock.

Marble A metamorphosed, recrystallized limestone.

Massive rock Said of a rock if the strength of the bond across partings or joints is comparable to the rock strength.

Metamorphic rocks Rocks which have been altered considerably by heat, pressure, and other factors.

Modulus of bending Rate of maximum fiber stress to maximum strain within the elastic limit of a stress-strain diagram obtained in a flexure test.

Modulus of elasticity Rate of change of strain as a function of stress.

Modulus of rigidity Rate of change of strain as a function of stress in a specimen subjected to shear or torsion loading.

Mohr's circle A graphical construction showing the state of stress at a point in a material.

Mohr's failure Failure of a material by fracture or permanent deformation when the shear stress on a real or potential fracture or slip plane has increased to a value that depends on the stress acting normal to the plane, or when the maximum principal stress has reached a certain limiting value.

Mylonite A fine-grained, laminated rock formed by extreme microbrecciation and milling (grinding) of rocks during movement on fault surfaces. Mylonitic seams may be very thin and difficult to find.

Orogeny The process of forming mountains, particularly by folding and thrusting. A major disturbance or mountain-building movement in the earth's crust.

Outcrop An actual exposure of rock (bedrock) or stratum projecting through the overlying cover of soil (the overburden soil removed).

Overburden The mantle of earth overlying rock formations.

Perched (or false) water table Caused by a layer of impervious material preventing the water from percolating to its natural level.

Permanent set Extent to which a material is permanently deformed by a specified load.

Permeability The flow per unit area under unit hydraulic gradient. DARCY's law of permeability: $v = ki$, where k = coefficient of permeability.

Petrology The study of rocks.

Pillar In-situ rock support between multiple openings.

Plastic Yielding or flowing under steady load. The term "plastic" is frequently used to describe processes involving yield and ductility. The term is used here in the sense defined in the mathematical theory of plasticity.

Plastic deformation Deformation that remains after the load causing it is removed. It is the permanent part of the deformation beyond the elastic limit of a material. It is also called plastic strain and plastic flow. Irrecoverable deformation.

Plasticity The property of a material that enables it to undergo permanent deformation without appreciable volume change or elastic rebound and without rupture.

Plutonic rock An igneous rock that has formed beneath the surface of the earth at great depth by consolidation from magma.

Poisson's ratio Ratio of lateral strain to axial strain in an axially loaded specimen.

Proportional limit The greatest stress at which a material is capable of sustaining load without deviation from proportionality of stress and strain (HOOKE's law).

Quartzite	A hard, granulose, metamorphic rock composed essentially of quartz. Sandstone cemented by silica.
Quasi-elastic	Quasi: as if, as though it were. Seemingly almost elastic.
Relaxation	Rate of reduction of stress in a material due to creep. An alternate term is stress relaxation.
Resilience	Measure of recoverable elastic energy in a deformed material. It is the amount of energy released when a load is removed from a specimen.
Rheology	Science dealing with the flow or deformation of matter. The study of time-dependent strain in both solids and liquids. Sometimes rheology describes the study of all types of deformation — elastic, plastic, and viscous, particularly the plastic flow of solids and flow of non-Newtonian liquids.
Rigidity	Lack of flexibility. The property of stiffness or of not yielding.
Rise of an arch	The maximum height that the undersurface of an arch rises.
Rock burst	A sudden and often violent failure of masses of rock in quarries, tunnels, and other underground openings.
Rockfall	The relatively free falling of a newly detached segment of bedrock of any size from a cliff, steep slope, cave, or arch.
Rock, massive	See Massive rock.
Rock slide	The downward and usually rapid movement of newly detached segments of the bedrock sliding on bedding, joint, or fault surfaces or any other plane of separation.
Roof	Top or ceiling of an underground excavation.
Rotational slide	See Slip.
Rupture	Failure of a material with development of fractures or shears.
Sandstone	Sedimentary rock composed of cemented sand grains, usually quartz.
Schist	A medium or coarse-grained metamorphic rock with subparallel orientation of the micaceous minerals that dominate its composition.

Sedimentary rocks Rocks formed by the accumulation of sediment derived from the breakdown of earlier rocks, by chemical precipitation, or by organic activity.

Self-supporting system A system in which the structural stresses are carried on the walls, pillars, and other unexcavated parts of the opening rather than on lining, packs, struts, braces, and chucks placed within the opening.

Semi-elastic Incompletely elastic.

Shaft An excavation (vertical) of limited area and comparatively great depth.

Shale A naturally densified, laminated sedimentary rock consisting predominantly of silt and clay. Shale splits easily parallel to the bedding.

Shearing An action of applied shear force causing or tending to cause two contiguous parts of a material to slide relative to each other in a direction parallel to their plane of contact resulting in a displacement of adjacent elements along a flat or curved surface.

Shear strength A measure of the resistance of a material to shear stress.

Shear stress A stress applied parallel to a surface or area. Maximum shear stress that can be sustained by a material before rupture. Applied stress that brings about rock and soil particles to slide on one another.

Silica The chemical composition of oxygen and silicon, which are the two most common elements in the earth's crust. Silicon dioxide: SiO_2.

Silicon A tetravalent, nonmetallic element (Si) that occurs combined as the most abundant element next to oxygen in the earth's crust and is used especially in alloys.

Silicones Resinous materials derived from organosiloxane polymers.

sine qua non (without which not) An absolutely indispensable or essential thing.

Slate A metamorphosed clay rock with a pronounced cleavage along which it readily splits.

Slip A relative displacement of points on opposite sides of a fault, measured on the surface of the fault. A minor fault. A slide.

Slip lines Orthogonal curves, whose directions at any point bisect the angles between the principal axes at that point.

Spring line Place on the side of a tunnel where the tunnel starts curving into the arch.

Stiffness The ratio of change of force to the corresponding change in deflection of an elastic element ($F \cdot L^{-1}$).

Strain Change in length per unit length (relative deformation).

Strain hardening Increase in strength caused by plastic deformation in cyclic loading.

Stratum A layer or bed of rock or soil.

Strength The maximum stress that a material can withstand during a normal short-time experiment without failing by rupture or continuous deformation. Rupture strength or breaking strength refers to the stress at the time of rupture.

Stress Force per unit area.

Stress field — biaxial One where the medium is subjected to a compressive or tensile stress in two directions (dimensions).

Stress field — hydrostatic One where the medium is subjected to equal stresses in three mutually perpendicular directions.

Stress field — unidirectional (unconfined) One where the medium is subjected to compressive or tensile stresses in one direction.

Stress relaxation Decrease in stress in a material subjected to prolonged constant strain at a constant temperature. Stress relaxation behavior is determined in a creep test.

Stress-strain diagram Graph of stress as a function of strain.

Stress trajectories (or isostatics) An orthogonal system of curves whose directions at any point are the directions of the principal axes. They, therefore, intersect a free boundary at right angles.

Strike Strike describes the geometry of attitude of inclined beds. When an inclined layer of unfolded or folded rock intersects a horizontal surface (land or water), the compass direction of the line formed by its intersection is called the strike. The dip of a fold is the acute angle between the horizontal and the axial plane, measured perpendicularly to the strike.

Structural bond A bond that joins basic, load-bearing parts of an assembly. The load may be static or dynamic.

Subsurface structure Any excavated or natural subsurface opening, or system of openings that is virtually self-supporting.

Syncline A trough-like fold (down-fold) in the rocks, with

	the beds dipping inward on either side.
System	A particular collection of matter which is being studied. An entity subject to study. An ordered grouping of certain elements and/or facts in a field of knowledge according to certain principles.
Tectonic	Of, pertaining to, or designating the rock structure and external forms resulting from the deformation of the earth's crust by deep-seated crustal and sub-crustal forces in the earth.
Tectonic creep	Fault creep of tectonic origin; also called slippage.
Tectonics	The study of the broader structural features of the earth and their causes. The phenomena associated with rock deformation and rock structures generally.
Tectonism	Crustal instability.
Tensile strength	Ultimate strength of a material subjected to tensile loading.
Thermal stress	Internal stress, caused in part by uneven heating.
Torsional deformation	Angular displacement of a specimen caused by a specified torque in torsion test.
Torsion test	Method for determining behavior of materials subjected to twisting loads (torque).
Toughness	Extent to which a material absorbs energy without fracture.
Trap	Old name for lava flow.
Traprock	Basaltic lava fields. Any of various dark-colored, fine-grained igneous rocks such as basalts used in building of roads.
Triaxial compressive strength	A measure of the strength of a material under a directed stress when the material is confined by a fluid under pressure, or by an enclosing jacket of metal or other material.
Tunnel	A passage in rock (soil) open at both ends.
Ultimate strength	The greatest engineering stress developed in material before rupture.
Unconfined compressive strength	A measure of the strength of a material under a directed, uniaxial stress.
Underground opening	Natural cavities or man-made excavations under the surface of the earth.
Viscoelastic	A material or a condition where strain resulting from stress is partly elastic and partly viscous.

Viscosity The property of liquids and solids that causes them to resist instantaneous change of shape and to produce strain that is dependent on time and the magnitude of stress.

Volcanic Pertaining to volcanoes or any rocks associated with volcanic activity at or below the surface.

Yield The onset of plastic deformation.

Yield point Stress at which strain increases without accompanying increase in stress. Also, see Elastic limit.

Yield strength Stress at which strain increases without accompanying increase in stress. It is an indication of maximum stress that can be developed in a material without causing plastic deformation. It is the stress at which a material exhibits a specified permanent deformation, and is a practical approximation of elastic limit.

Young's modulus Alternate term for the modulus of elasticity in tension or compression. It is a constant for each elastic material that equals the ratio of stress to strain ($E = \sigma/\varepsilon$).

Zeta potential ζ **(cross potential)** Electric potential across the double layer or interface between the solid and liquid phases of a soil or rock systems.

APPENDIX 4

ROCK DEFECTS

Term	Description
1	2
Bedding joint	Joint parallel to bedding.
Bedding plane	In sedimentary or stratified rocks, the division planes which separate the individual layers, beds, or strata.
Crack	A break without parting. To crack: to break so that fissures appear on the surface.
Fault	The displacement of rocks along a zone of fracture.
Fissure	An extensive crack, break, or fracture in the rocks.
Fold	A flexure or bend in strata or any planar structure when rocks are in a plastic condition.
Fracture	The manner of breaking and appearance of a mineral when broken. Breaks in rocks due to intense folding of faulting.
Joint	Fracture in rock, generally more or less vertical or transverse to bedding, along which no appreciable movement has occurred.

Joint set	A group of more or less parallel joints.
Parting	A small joint in coal or rock. A thin depositional layer separating thick deposits, as shale in a coal seam; also a joint or fissure.
Seam	A quarryman's term for a joint, cleft, or fissure in rocks. In mining and geology, the term designates a thin layer of stratum, usually of coal, clay or other material.

APPENDIX 5

CONVERSION FACTORS FOR UNITS OF MEASUREMENT

A. Use of SI Units in Geotechnical Engeneering

B. Conversion Factors

A. USE OF SI UNITS IN GEOTECHNICAL ENGINEERING

acceleration, *a*	m/s²
acceleration of gravity, *g*	9.80665 m/s²
velocity, *v* coefficient of permeability, *k*	m/s cm/s
length, *L*; depth *L*; *d* displacement; settlement, *s*; deflection, *ΔL* distance, *L*	m; cm; mm cm km; m; cm
area, *A*	cm²; m²
volume, *V*	cm³; m³
section modulus, *Z*	cm³; m³
moment of inertia, *I*, *J*	cm⁴; m⁴
forces of all kinds	N

force per unit length, P; surcharge, p	N/m
modulus of elasticity, E	N/m², or in Pa
allowable pressure on soil and rock, σ_{all}	N/m², or in Pa
contact pressure on soil, σ_o	N/m², or in Pa
stress in soil, σ; surcharge, p, q	N/m², or in Pa
shear stresses, τ, s; cohesion c	N/m², or in Pa
mantle (skin) friction; adhesion f	N/m², or in Pa
subgrade reaction of soil, C	N/cm³
unit weight of soil, γ	N/m³
total earth pressures and resistances, E_a, E_o, E_p	N/m or N

B. CONVERSION FACTORS

LENGTH

1 mm	= 0.03937 in. 1 cm = 10 mm = 0.3937 in.
1 m	= 100 cm = 1000 mm = 39.37 in. = 3.28083 ft.
1 in.	= 2.54 cm 1 ft = 0.304801 m 1 yd = 0.91443 m

AREA

1 cm²	= 0.15500 in.² = 0.001076387 ft² = 0.0001 m² = 100 m²
1 m²	= 10.76387 ft² = 10^{-4} ha = 10^4 cm²
1 ha	= 2.4710624 acres = 10,000 m² = 100 ares
1 in.²	= 6.45164 cm² = 6.45164×10^{-4} m²
1 ft²	= 0.0929 m² = 2.296×10^{-5}acre = 144 in.²
1 acre	= 43,560 ft² = 40,468420 ares = 0.4046842 ha
1 are	= 100 m²

VOLUME

1 cm³	= 10^{-6} m³ = 0.06102 in.³ = 3.531×10^{-5} ft³
1 m³	= 10^6 cm³ = 6.102×10^4 in.³ = 35.31445 ft³
1 in.³	= 5.787×10^{-4} ft³ = 16.38716 cm³ = 1.6387×10^{-5} m³
1 ft³	= 1728 in.³ = 2.832×10^4 cm³ = 0.02832 m³

CAPACITY

1 liter	= 1000 cm^3 = 0.001 m^3 = 61.02 $in.^3$ = 3.531 × 10^{-2} ft^3
	= 0.26417 U.S. gallons
1 U.S. gallon	= 231 $in.^3$ = 0.1337 ft^3 = 3785 cm^3 = 3.78543 liters
	= 0.00379 m^3
1 cm^3	= 0.001 liters = 2.642 × 10^{-4} gallons
1 m^3	= 1000 liters = 264.17047 gallons = 35.31 ft^3
1 ft^3	= 7.481 gallons = 28.32 liters

FORCE

Force unit: newton (N). Formula: (kg) (m/s^2)

Conversion factors for force apply only under standard acceleration of 9.80665 m/s^2 due to gravity.

Notations:

N	—	newton
Pa	—	pascal
g_f	—	force gram
kg_f	—	force kilogram
t_f	—	metric force ton
ton	—	short ton = 2000 lb
k	—	kip = 1000 lb

1 N	=	(1.00000) (10^5)	dynes
	=	(1.01971) (10^2)	g_f
	=	(1.01971) (10^{-1})	kg_f
	=	(1.01971) (10^{-4})	t_f
	=	(2.24809) (10^{-1})	lb
	=	(2.24809) (10^{-4})	k
	=	(1.12405) (10^{-4})	ton
1 kN	=	(1.00000) (10^3)	N
	=	(1.00000) (10^{-3})	MN
	=	(1.00000) (10^8)	dynes
	=	(1.01971) (10^5)	g_f
	=	(1.01971) (10^2)	kg_f
	=	(1.01971) (10^{-1})	t_f
	=	(2.24809) (10^2)	lb
	=	(2.24809) (10^{-1})	k
	=	(1.12405) (10^{-1})	ton

1 MN	=	$(1.00000)(10^{6})$	N
	=	$(1.00000)(10^{3})$	kN
	=	$(1.01971)(10^{8})$	g_f
	=	$(1.01971)(10^{5})$	kg_f
	=	$(1.01971)(10^{2})$	t_f
	=	$(2.24809)(10^{5})$	lb
	=	$(2.24809)(10^{2})$	k
	=	$(1.12405)(10^{2})$	ton
1 dyne	=	$(1.00000)(10^{-5})$	N
	=	$(1.01971)(10^{-3})$	g_f
	=	$(1.01971)(10^{-6})$	kg_f
	=	$(1.01971)(10^{-9})$	t_f
	=	$(2.24809)(10^{-6})$	lb
	=	1.00000	erg/cm
1 g_f	=	$(9.80665)(10^{-3})$	N
	=	$(9.80665)(10^{2})$	dynes
	=	$(1.00000)(10^{-3})$	kg_f
	=	$(1.00000)(10^{-6})$	t_f
	=	$(2.20462)(10^{-3})$	lb
	=	$(1.10231)(10^{-6})$	ton
1 kg_f	=	9.80665	N
	=	$(9.80665)(10^{5})$	dynes
	=	$(1.00000)(10^{3})$	g_f
	=	$(1.00000)(10^{-3})$	t_f
	=	2.20462	lb
	=	$(2.20462)(10^{-3})$	k
	=	$(1.10231)(10^{-3})$	ton
1 t_f	=	$(9.80665)(10^{3})$	N
	=	$(1.00000)(10^{8})$	dynes
	=	$(1.00000)(10^{6})$	g_f
	=	$(1.00000)(10^{3})$	kg_f
	=	$(2.20462)(10^{3})$	lb
	=	2.20462	k
	=	1.10231	ton
1 lb	=	4.44822	N
	=	$(4.44822)(10^{5})$	dynes
	=	453.592	g_f
	=	0.453592	kg_f
	=	$(4.53592)(10^{-4})$	t_f
	=	$(1.00000)(10^{-3})$	k
	=	$(5.00000)(10^{-4})$	ton

1 k (= kip)	=	(4.44822) (10^3)	N
	=	(4.44822) (10^8)	dynes
	=	(4.53592) (10^5)	g_f
	=	(4.53592) (10^2)	kg_f
	=	(4.53592) (10^{-1})	t_f
	=	(1.00000) (10^3)	lb
	=	(5.00000) (10^{-1})	ton
1 ton	=	(8.89644) (10^3)	N
	=	(8.89644) (10^8)	dynes
	=	(9.07184) (10^5)	g_f
	=	(9.07184) (10^2)	kg_f
	=	(9.07184) (10^{-1})	t_f
	=	(2.00000) (10^3)	lb
	=	2.00000	k

STRESS

1 N/m^2 = 1 Pa	=	(1.00000) (10^{-4})	N/cm^2
	=	(1.01972) (10^{-2})	g_f/cm^2
	=	(1.01972) (10^{-5})	kg_f/cm^2
	=	(1.01972) (10^{-1})	kg_f/m^2
	=	(1.01972) (10^{-4})	t_f/m^2
	=	(1.45038) (10^{-4})	lb/in.^{2}
	=	(2.08854) (10^{-2})	lb/ft^2
	=	(2.08854) (10^{-5})	k/ft^2
	=	(1.04427) (10^{-5})	ton/ft^2
1 kg_f/cm^2	=	(9.80665) (10^4)	N/m^2
	=	9.80665	N/cm^2
	=	(1.00000) (10^4)	kg_f/m^2
	=	(1.00000) (10)	t_f/m^2
	=	(1.42233) (10)	lb/in.^{2}
	=	(2.04817) (10^3)	lb/ft^2
	=	2.04817	k/ft^2
	=	1.02409	ton/ft^2
1 t_f/m^2	=	(9.80665) (10^3)	N/m^2
	=	(1.00000) (10^3)	kg_f/m^2
	=	(1.00000) (10^{-1})	kg_f/cm^2
	=	(1.02408) (10^{-1})	ton/ft^2
	=	(2.04817) (10^{-1})	k/ft^2
	=	(2.04817) (10^2)	lb/ft^2
	=	1.42233	psi

1 lb/in.^{2}	=	(6.89476) (10^3)	N/m^2
	=	(6.79476) (10^{-1})	N/cm^2
	=	(7.03069) (10)	g_f/cm^2
	=	(7.03069) (10^{-2})	kg_f/cm^2
	=	(7.03069) (10^2)	kg_f/m^2
	=	(7.03069) (10^{-1})	t_f/m^2
	=	(1.44000) (10^2)	lb/ft^2
	=	(1.44000) (10^{-1})	k/ft^2
	=	(7.20000) (10^{-2})	ton/ft^2
1 lb/ft^2	=	(4.78803) (10)	N/m^2
	=	(4.78803) (10^{-3})	N/cm^2
	=	(4.88242) (10^{-1})	g_f/cm^2
	=	(4.88242) (10^{-4})	kg_f/cm^2
	=	4.88242	kg_f/m^2
	=	(4.88242) (10^{-3})	t_f/m^2
	=	(6.94444) (10^{-3})	lb/in.^{2}
	=	(1.00000) (10^{-3})	k/ft^2
	=	(5.00000) (10^{-4})	ton/ft^2
1 k/ft^2	=	1000	lb/ft^2
	=	0.500000	ton/ft^2
	=	(4.78803) (10^4)	N/m^2
	=	(4.88242) (10^{-1})	kg_f/cm^2
	=	(4.88242) (10^3)	kg_f/m^2
	=	4.88242	t_f/m^2
	=	6.94444	psi
	=	(1.00000) (10^{-6})	lb/ft^2
	=	(5.00000) (10^{-1})	ton/ft^2
1 ton/ft^2	=	2000	lb/ft^2
	=	(9.57605) (10^4)	N/m^2
	=	(9.76485) (10^{-1})	kg_f/cm^2
	=	(9.76485) (10^3)	kg_f/m^2
	=	9.76485	t_f/m^2
	=	(1.38888) (10)	psi
	=	(2.00000) (10^3)	lb/ft^2
	=	2.00000	k/ft^2

UNIT WEIGHT

1 N/m^3	=	(1.01972) (10^{-4})	g_f/cm^3
	=	(1.01972) (10^{-1})	kg_f/m^3
	=	(1.01972) (10^{-4})	t_f/m^3

	=	$(6.36586)(10^{-3})$	lb/ft^3
	=	$(3.18293)(10^{-6})$	ton/ft^3
1 g_f/cm^3	=	$(9.80665)(10^3)$	N/m^3
	=	$(1.00000)(10^{-3})$	kg$_f$/cm^3
	=	$(1.00000)(10^3)$	kg$_f$/m^3
	=	1.00000	t$_f$/m^3
	=	$(6.24277)(10)$	lb/ft^3
	=	$(3.12139)(10^{-2})$	ton/ft^3
1 kg$_f$/m^3	=	9.80665	N/m^3
	=	$(1.00000)(10^{-3})$	g$_f$/cm^3
	=	$(1.00000)(10^{-6})$	kg$_f$/cm^3
	=	$(1.00000)(10^{-3})$	t$_f$/m^3
	=	$(6.24277)(10^{-2})$	lb/ft^3
	=	$(3.12139)(10^{-5})$	ton/ft^3
1 t$_f$/m^3	=	$(9.80665)(10^3)$	N/m^3
	=	$(1.00000)(10^3)$	kg$_f$/m^3
	=	$(6.24277)(10)$	lb/ft^3
	=	$(3.12139)(10^{-2})$	ton/ft^3
1 lb/ft^3	=	$(1.57092)(10^2)$	N/m^3
	=	$(1.60189)(10)$	kg$_f$/m^3
	=	$(1.60189)(10^{-2})$	t$_f$/m^3
	=	$(5.00000)(10^{-4})$	ton/ft^3
1 ton/ft^3	=	$(3.14184)(10^5)$	N/m^3
	=	$(3.20379)(10^4)$	kg$_f$/m^3
	=	$(3.20379)(10)$	t$_f$/m^3
	=	$(2.00000)(10^3)$	lb/ft^3

TEMPERATURE

$$1\,[K] = 1\,[C] = (1.8)\,[F]$$

$$1\,[F] = (0.555)\,[C] = (0.555)\,[K]$$

Absolute zero: 0 °K = –273.15 °C = –459.67 °F

$$\begin{aligned} T_C &= (5/9)\,(T_F - 32°) \\ &= T_K - 273.15° \\ T_K &= T_C + 273.15° \\ &= (T_F + 459.67°)/1.8 \\ T_F &= (9/5)\,(T_C) + 32° \\ &= (T_C + 17.8°)\,(1.8) \end{aligned}$$

HEAT

1 BTU = 1 B = (1.05435) (10^3) Nm (or Joule, J.)
= 0.252 Cal = 252 cal

1 Cal = 1000 cal = 3.968 B

1 cal = 0.003968 B

1 Nm = 1 J = 1 ws
= (2.77778) (10^{-4}) wh
= (2.77778) (10^{-7}) kwh
= (2.38846) (10^{-4}) Cal
= (2.38846) (10^{-1}) cal
= (9.47741) (10^{-4}) B
= (3.77672) (10^{-7}) PSh
= (3.72505) (10^{-7}) HPh

1 Cal/(m) (h) (°C) = (0.67197) B/(ft) (h) (°F)
= (8.064) B-in./(ft^2) (h) (°F)
= (2.778) (10^{-3}) cal/(cm) (s) (°C)
= (1.163) (10^{-2}) w/(cm) (°C)

1 cal/(cm) (s) (°C) = (360) Cal/(m) (h) (°C)
= (241.9) B/(ft) (h) (°F)
= (2.9) (10^3) B-in./(ft^2) (h) (°F)
= (4.18680) w/(cm) (°C)

1 B/(ft) (h) (°F) = (1.48817) Cal/(m) (h) (°C)
= (4.134) (10^{-3}) cal/(cm) (s) (°C)

	=	(12) B-in./(ft^2) (h) (°F)
	=	(1.730) (10^{-2}) w/(cm) (°C)
1 B/(ft^2) (h) (°F/in.)	=	(0.124) Cal/(m) (h) (°C)
	=	(3.447) (10^{-4}) cal/(cm) (s) (°C)
	=	(8.34) (10^{-2}) B/(ft) (h) (°F)
	=	(1.44314) (10^{-3}) w/(cm) (°C)
1 w/(cm) (°C)	=	(85.9845) Cal/(m) (h) (°C)
	=	(0.2388) cal/(cm) (s) (°C)
	=	(57.81477) B/(ft) (h) (°F)
	=	(6.92933) B/(ft^2) (h) (°F/in.)

ABOUT THE AUTHOR

Alfreds Richards Jumikis was born in Riga, Latvia. Since 1952, Dr. Jumikis has been a member of the Department of Civil and Environmental Engineering of the College of Engineering of Rutgers, The State University of New Jersey. Professor Jumikis had previously taught at the University of Delaware, and at the University of Latvia in Riga. He holds earned academic degrees of Dr. Eng. Sc., Dr. techn., and Dr.-Ing. from the University of Latvia, the Technical University in Vienna, and the University of Stuttgart, respectively.

Dr. Jumikis is the author of more than one hundred publications, among them 17 books and booklets on shale rocks, soil mechanics, foundation engineering, thermal soil mechanics, and a book on Glossary of Terms in Thermal Soil Mechanics.

Professor Jumikis has lectured and presented papers at national and international conferences.

Dr. Jumikis is a member of several professional engineering and scientific societies, and is the recipient of three National Science Foundation grants for his research in thermal soil mechanics.

Alfreds R. Jumikis is a registered professional engineer (P. E.), and consults on problems in geotechnics.

INDEX

NAME INDEX

SUBJECT INDEX

D

G

H

I

J

K

L

M

O

P

T